Textbook of
Physics for Engineers

Textbook of Physics for Engineers

Suresh Chandra
Mohit K. Sharma
Monika Sharma

Alpha Science International Ltd.
Oxford, U.K.

Textbook of Physics for Engineers
310 pgs. | 86 figs. | 02 tbls.

Suresh Chandra
Department of Physics
Lovely School of Sciences
Lovely Professonal University
Phagwara, Punjab

Mohit K. Sharma
Monika Sharma
School of Studies in Physics
Jiwaji University
Gwalior

Copyright © 2015

ALPHA SCIENCE INTERNATIONAL LTD.
7200 The Quorum, Oxford Business Park North
Garsington Road, Oxford OX4 2JZ, U.K.

www.alphasci.com

Printed from the camera-ready copy provided by the Authors.

ISBN 978-1-84265-941-0

***Dedicated to**
my Maternal Grand Parents
Shri Naubat Ram Sharma & Smt. Ganga Devi Sharma*

Suresh Chandra

Preface

Physics is an important aspect for all branches of technology. One may undoubtedly mention that the technology is an applied aspect of Physics. Thus, for a strong basis of technology, knowledge of the basics of Physics is essentially required. Though at school level, Physics is taught in detail, but, at that time, most of the students do not realize the importance of Physics for understanding each branch of technology. Keeping in view that aspect, undergraduate engineering students are taught some important parts of Physics.

A number of books are written for B. Tech. students. Many of them claim their books cover a complete course. We also decided to present this book in a student friendly style and simple language. The material presented in this book is based on long and extensive teaching and research experience of Prof. Suresh Chandra in various universities over a period of 38 years. The experience of Prof. Suresh Chandra has been compiled in the form of this book by Mohit K. Sharma and Monika Sharma. The text is supplemented by exercises and multiple choice questions, which help in proper understanding of the subject. For further improvement, scope is always there. We shall appreciate receiving comments/suggestions from the readers of this book, which can be sent at the following emails also:

suresh492000@yahoo.co.in
mohitkumarsharma32@yahoo.in
monika3273@yahoo.in

While preparing the manuscript, we have been helped, advised, and encouraged by our seniors, friends, and colleagues working in various institutions, and by our friends in personal life. We are thankful to all of them. Suresh Chandra is thankful to the authorities of the Lovely Professional University for encouragement. He is specially thankful to his wife Mrs. Purnima Sharma for her valuable cooperation in his life and for sharing a major part of responsibility of family affairs. We are deeply grateful to our family members who always have been source of inspiration and happiness for us. Last but not the least, we are highly thankful to the Publisher.

Suresh Chandra
Mohit K. Sharma
Monika Sharma

Contents

I. Measurements and Errors

1. Physical quantities

Let us consider a body. The body may have length. breadth, height, volume, mass and density. When the body is in motion, we can measure its distance, velocity, acceleration, momentum, angular momentum, energy. The body may be affected by applying some external force(s). Such force may be a mechanical force, electric force, magnetic force, gravitational force. There may be others parameters also, such as temperature, related to the body. All these parameters pertaining to a body are known as the physical quantities. The physical quantities may be classified into two categories: (i) Fundamental quantities and (ii) Derived quantities.

1.1 Fundamental quantities

The fundamental quantities are those which cannot be expressed in terms of others. They serve as the basic constants of measurement. Commonly used fundamental quantities are the length, mass and time. (Temperature is also considered as a fundamental quantity.) In the SI (standard of international) system, the units of the fundamental quantities are given in the following table.

Quantity	Unit	Symbol
Length	meter (m)	L
Mass	kilogram (kg)	M
Time	second (s)	T

1.2 Derived quantities

The derived quantities are those which can be expressed in terms of fundamental quantities. Some examples of the derived quantities are the area, volume, density, velocity, acceleration, mcmentum, pressure, force etc. Information about some derived quantities is given in the following table.

Quantity	Expression	Dimensional formula
Area	length × breadth	L^2
Volume	length × breadth × height	L^3
Density	mass/volume	ML^{-3}
Velocity	distance/time	LT^{-1}
Acceleration	change in velocity/time	LT^{-2}
Force	mass × acceleration	MLT^{-2}
Pressure	force/area	$ML^{-1}T^{-2}$
Momentum	mass × velocity	MLT^{-1}

2. Least count

For measurement of physical quantities, various apparatuses (instruments) are used. Each of the instruments has some limitations. For example, a meter-scale can measure the length of an object in millimeters. The length may be, for example, 22 mm, 23 mm and so on. But, it cannot measure, for example, 22.4 mm, 24.43 mm, 32.345 mm. It shows the limitation of the meter-scale, as it cannot measure a fraction of millimeter. On the other side, a vernier calliper can measure 22.4 mm. But, it cannot measure 24.43 mm, 32.345 mm (for example), as it can measure one-tenth of millimeter. It shows the limitation of the vernier calliper. The screw gauge can measure 24.43 mm. But, it cannot measure 32.345 mm (for example), as it can measure one-hundredth of millimeter. The minimum difference which can be measured by an apparatus (instrument) is known as the least count of the apparatus (instrument). In the aforesaid discussion, the least count of meter-scale is 1 mm, of vernier calliper is 0.1 mm, of screw gauge is 0.01 mm.

Every apparatus has a least count. The least count of a wrist watch is 1 second.

Exercise 1. When measured with very accurate apparatus, length of a small object is found to be 1.34567 cm. What will be the length of this object when it is measured with the help of: (i) vernier calliper and (ii) screw gauge.

Solution: (i) The least count of vernier calliper is 0.1 mm (0.01cm) and therefore the value of the length of the object measured with the help of the vernier calliper will be 1.35 cm.

(ii) The least count of screw gauge is 0.01 mm (0.001cm) and therefore the value of the length of the object measured with the help of the screw gauge will be 1.346 cm.

3. Uncertainty

The maximum uncertainty in a measurement taken with the help of an apparatus is half of its least count.[1]

$$\text{Maximum uncertainty} = \frac{\text{Least count}}{2}$$

For example, let us consider that the length of a wooden block is measured with the help of vernier calliper having least count 0.1 mm. The maximum uncertainty in the measurement would be

$$\text{Maximum uncertainty} = \frac{\text{Least count}}{2} = \frac{0.1}{2} = 0.05 \text{ mm}$$

It can be visualize in the following manner. Suppose the length of a wooden block measured with the help of vernier calliper is 12.4 mm. The actual length may vary from 12.35 mm to 12.49999... mm. It shows that there is uncertainty of ± 0.05 mm in the measurement with the help of vernier calliper.

Exercise 2. Measurement is taken with the help of a screw gauge having least count 0.01 mm. Calculate the maximum possible uncertainty in the measurement.

Solution: The maximum possible uncertainty in the measurement is

$$\text{Maximum uncertainty} = \frac{\text{Least count}}{2} = \frac{0.01}{2} = 0.005 \text{ mm}$$

Exercise 3. Diameter of a wire measured with the help of screw gauge is found to be 3.45 mm. What is the maximum probable error in this measurement.

Solution: The value is measured up to second place of millimeter. Thus, the least count of screw gauge is 0.01 mm. Therefore, the maximum probable error in the measurement is

$$\text{Maximum probable error} = \frac{0.01}{2} = 0.005 \text{ mm}$$

4. Accuracy

In order to understand about the accuracy of measurements, let consider the following example. Suppose, the length of a small size object can be

---

[1]Some one may have an opinion that the uncertainty in a measurement is equal to its least count.

measured with the help of: (i) vernier calliper and (ii) screw gauge. The least count of vernier calliper is 0.1 mm whereas that of the screw gauge is 0.01 mm. Obviously, the length measured with the help of vernier calliper will be reported up to one decimal place in millimeter. When the length is measured with the help of screw gauge, the value will be reported up to second decimal place in millimeter. Thus, the accuracy of the measurement made with the help of the screw gauge is better than that made with the help of vernier calliper. (When the size of the object is large, the measurement cannot be made with the help of screw gauge.) When two different instruments are used for measurement of a common object, the accuracy of measurement is decided on the basis of the least counts of the instruments.

Exercise 4. Length of a small piece is measured with the help of (i) meter-scale, (ii) vernier calliper, and (iii) screw gauge. Which measurement will be the most accurate among them.

Solution: The least counts of meter-scale, vernier calliper, screw gauge are, respectively, 1 mm, 0.1 mm, 0.01 mm. As the object is common, the measurement made with the help of the screw gauge would be most accurate.

Exercise 5. For measuring masses of small objects, we have two balances A and b having least counts 0.01 gm and 0.0001 gm, respectively. Measurement of which balance will be more accurate.

Solution: As the least counts of the balance B is better than that of the balance, A, the balance B will measure the mass more accurately.

5. Resolution

When we can distinguish between two objects, the situation is said to be resolved. Suppose, we have two objects and we can differentiate between their lengths, their lengths are said to be resolved. On the other side, when it is not possible to measure difference between lengths of two objects, their lengths are said to be unresolved. When one used instrument with better least count the unresolved situation may convert to the resolved situation. For example, let us consider two objects A and B having lengths 5.32 cm and 5.34 cm, respectively. The lengths of these objects cannot be resolved when the measurements are made with the help of a meter-scale, as the least count of the meter-scale is 0.1 cm. The meter-scale will measure the lengths of both A and B as 5.3 cm. The lengths of these objects can be resolved when the measurements are made with the help of a vernier calliper, as the least count of the vernier calliper is 0.01 cm. The vernier calliper will measure the lengths of A and B as 5.32 cm and 5.34 cm, respectively.

Exercise 6. Least counts of meter-scale, vernier calliper, screw gauge are 1 mm, 0.1 mm, 0.01 mm, respectively. Two small objects A and B have lengths 5.42 mm and 5.44 mm, respectively. Answer with reason which of the meter-scale, vernier calliper, screw gauge can resolve the lengths of the objects.

Solution: The meter-scale has least count of 1 mm and therefore will measure the lengths of both the objects A and B as 5 mm. Thus, the meter-scale can not resolve the lengths of the objects.

The vernier calliper has least count of 0.1 mm and therefore will measure the lengths of both the objects A and B as 5.4 mm. Thus, the vernier calliper can not resolve the lengths of the objects.

The screw gauge has least count of 0.01 mm and therefore will measure the lengths of the objects A and B as 5.42 mm and 5.44 mm, respectively. Thus, the screw gauge will resolve the lengths of the objects.

6. Random and systematic errors

Though a better way to estimate uncertainty in a measurement is to make multiple measurements of the same quantity, but not all types of experimental uncertainties can be assessed with the help of repeated measurements. For this reason, uncertainties are classified into two categories: (i) random errors and (ii) systematic errors.

6.1 Random errors

The random errors can be treated statistically. That is, we take observations of a physical quantity a number of times and then take arithmetic mean of these observations. Let us consider an example of simple pendulum where we measure the time period (time required for one oscillation) of a steadily moving pendulum. Here, one source of error will be our reaction time in starting and stopping the stop-watch. If our reaction times are always exactly the same, these two errors would cancel to each other. In practice, our reaction time may vary each time. We may, for example, delay in starting the stop-watch, and so underestimate the time, or we may delay in stopping the stop-watch, and so overestimate the time. We may start or stop the stop-watch early. Since either possibility is likely, both the size and the sign of effect are random.

If we repeat the measurement several times, our variable, the reaction time will show a variation in the measured times. By looking at the spread in the measured times, we can get a reliable estimate of this kind of error.

6.2 Systematic errors

The systematic errors cannot be estimated statistically. It is due to error in the apparatus being used for measurement. In the example of simple pendulum, when our stop-watch is running consistently slow, then all our times will be underestimated, and no amount of repetition (with the same watch) will estimate this kind of error. This kind of error is known as the systematic error, because it always pushes our result in the same direction. Systematic errors cannot be discovered with the help of statistical analysis. They are often hard to evaluate or even to detect. The only way to eliminate the systematic error is to replace the apparatus by an accurate one. In our experiments, we often (but not always) assume that systematic errors have been made much smaller than the required precision. That is, we assume that the apparatus is working properly.

7. Significant figures

In the number system with base 10 (called the decimal system), a number is a composition of ten digits 0, 1, 2, 3, 4, 5, 6, 7, 8, and 9. In the decimal system, we generally deal with two types of numbers: (i) integers, and (ii) real numbers. An integer is a signed or unsigned whole numbers without a decimal point or fraction. Examples of integers are: 15, 140, −752, −6859 etc. A real number in general has a decimal point. It is the decimal point which helps in distinguishing between a real number and an integer. For example, 47 is an integer and 47. (along with the decimal point) is a real number. The real numbers may be expressed either in the fractional form or in the exponential form (also known as the scientific notation). Some examples of real numbers are: 34.5700, −67.04, 4.63×10^2, 5.87×10^{-2}, -6.15×10^3, -6.704×10^{-8}, etc. Here, the first two numbers are expressed in the fractional form whereas the next four numbers are expressed in the exponential form.

Whether all the digits in a given number are significant or not, let us look into it. A significant figure or significant digit in a number is any one of the digits 1, 2, 3, ..., 9. The digit 0 is a significant figure except when it is used to fix the decimal point or to fill the places of unknown or discarded digits. For example, in the number 0.00574, the significant figures are 5, 7, and 4. Here, the zeroes are not significant figures. In the number 5609, all the digits including the zero, are significant figures. The ambiguity for deciding about the significant figures in a number can be sorted out by expressing a number in the scientific notation. Then the number of significant figures is indicated by the factor at the left. For example, in the number 436000, the number of significant figures may be ambiguous, whereas the numbers 4.36×10^5, 4.360×10^5, 4.3600×10^5 have three, four and five significant figures, respectively. The number 0.005097 can be expressed as 5.097×10^{-3}. Then the significant figures are 5, 0, 9, and 7. Furthermore, in the numbers 3.497×10^{-3}, 3.4970×10^{-3}, 3.49700×10^{-3}, the number of significant figures are four, five and six, respectively.

The following rules for determining the number of significant figures have been set up.

1. All non-zero digits are significant. For example, 148.352 has six significant figures.

2. All zeroes occurring between two non-zero digits are significant. For example, 402.01 has five significant figures.

3. All zeroes which are right of a decimal point and left of a non-zero digits are never significant. For example, 0.0056 has two significant figures; 0.000508 has three significant figures.

4. All zeroes which are right of a decimal point and right of a non-zero digit are significant. For example, 0.08500 has four significant figures; 50.000 has five significant figures.

5. All zeroes to the right of the last (right most) non-zero digit are not significant. For example, 20300 has three significant figures. However, zeroes to the right of the last non-zero digit are significant if they come from a measurement. Suppose, the distance between two cities is 200 m, then there are three significant figures.

Exercise 7: Find the significant figures in the numbers: 0.00608, 4.070, 2.03×10^2, 3.040×10^2, 2.0400×10^2, 4.16×10^{-2}, 2.460×10^{-2}, 4.3700×10^{-2}.

Solution: The significant figures in the numbers are given in the following table.

Number	Significant figures in the number
0.00608	6, 0, 8
4.070	4, 0, 7, 0
2.03×10^2	2, 0, 3
3.040×10^2	3, 0, 4, 0
2.0400×10^2	2, 0, 4, 0, 0
4.16×10^{-2}	4, 1, 6
2.460×10^{-2}	2, 4, 6, 0
4.3700×10^{-2}	4, 3, 7, 0, 0

8. Approximate numbers

A large number of numbers, we deal with, consists of approximate numbers. There are exact numbers also. The numbers such as 5, 6/7, 356, etc. are exact, as there is no approximation or uncertainty associated with them.[2] Though the numbers, such as π, $\sqrt{3}$, e, etc. are exact numbers, they however cannot be expressed exactly in the form of digits. When expressed in the digital form, they must be, for example, written as 3.1416, 1.7321, 2.7183, etc. These numbers are therefore only approximations to the true values[3] and in such cases are called the approximate numbers. An approximate number is therefore a number which is used as an approximation to its exact value and differs only slightly from the exact value for which it stands.

9. Rounding off numbers

In order to understand the process of rounding off the numbers, let us, for example, attempt to divide 22.7 by 14.3 to get

$$22.7/14.3 = 1.587412587....$$

Obviously, this number never terminates. For using such a number in computation, we would like to cut it down to a manageable form, such as 1.59, 1.587, 1.5874, ..., 1.587412587 etc. This process of retaining, as many as desired digits and cutting off less important digits is known as the rounding off of the number. After the rounding off, we obtain an approximate number. While rounding off a number, care is taken that it should cause the least possible error. In order to round off a number

---

[2]During the execution, the number 6/7 also becomes as an approximate number.
[3]We have $\pi = 3.141592654....$, $\sqrt{3} = 1.732050808....$, e $= 2.718281828....$

to an appropriate value, say, up to nth place, we discard all the digits to the right of the nth place according to the following rules:

(a) When the discarded number is less than half a unit in the nth place, leave the digit at the nth place unchanged. In this case the digit at the $(n+1)$th place is less than 5.

(b) When the discarded number is more than half a unit in the nth place, add 1 to the digit at the nth place. In this case the digit at the $(n+1)$th place is more than 5.

(c) When the discarded number is exactly half a unit in the nth place, leave the nth digit unchanged when it is an even number, but increase it by 1 when it is an odd number. Obviously, it will leave an even number in both cases at the nth place. In this case the digit at the $(n+1)$th place is 5.

According to the above rules, some numbers rounded off to five significant digits are given in the following table:

Number	Rounded number
43.57297	43.573
85.89344	85.893
5.47996001	5.4800
35.623479	35.623
27.0365	27.036
54.1275	54.128

When the number is rounded off, it is said to be correct to the nth place. Thus, by rounding off a number, some error is introduced. When the rounding off the numbers is done according to the said rules, in the computation of a mathematical expression, the errors due to rounding off may be largely canceled by one another.

Exercise 8: Round off the numbers 43.602356, -5.389234, 647.0025679, 809.093546, 812.8567 up to one, two and three decimal points.

Solution: The values rounded off up to one, two and three decimal points are given in the following table.

Number	Rounded up to one dec point	Rounded up to two dec points	Rounded up to three dec points
43.602356	43.6	43.60	43.602
-5.389234	-5.4	-5.39	-5.389
647.0025679	647.0	435.00	647.002
809.093546	809.1	809.09	809.094
812.8567	812.9	812.86	812.856

Exercise 9: Round off the numbers 134789.2, -578502, 309460, 476829, 9312.456 up to unit, ten and hundred positions.

Solution: The values rounded off up to unit, ten and hundred positions are given in the following table.

Number	Rounded up to unit position	Rounded up to ten position	Rounded up to hundred position
134789.2	134789	134790	134800
-578502	-578502	-578500	-578500
309460	309460	309460	309500
476829	476829	476830	476800
9312.456	9312	9310	9300

9.1 Truncation error

The error which arises due to truncation of a given number is known as the truncation error. Suppose we have a number 5.217. If we truncate it after second decimal place, *i.e.*, we get 5.21. The error introduced due to truncation is $5.217 - 5.21 = 0.007$. The percentage error introduced is

$$\frac{5.217 - 5.21}{5.217} \times 100 = 0.13\%$$

9.2 Round-off error

The error which arises due to rounding-off of a given number is known as the round-off error. Suppose we have a number 6.436. If we round-off it after second decimal place, *i.e.*, we get 6.44. The error introduced due to r rounding-off is $6.44 - 6.436 = 0.004$. The percentage error introduced is

$$\frac{6.44 - 6.436}{6.436} \times 100 = 0.062\%$$

Exercise 10: After calculations of an expression, we got a value 8.356. If we want to have the value up to first decimal place, calculate the (i) truncation error and (ii) rounding-off error.

Solution: (i) On truncation, we shall get the value 8.3. Then the error introduced is $8.356 - 8.3 = 0.056$—. Thus, the percentage truncation error introduced is

$$\frac{8.356 - 8.3}{8.356} \times 100 = 0.67\%$$

(ii) On rounding-off, we shall get the value 8.4. Then the error introduced is $8.4 - 8.356 = 0.044$. Thus, the percentage rounding-off error introduced is

$$\frac{8.4 - 8.356}{8.356} \times 100 = 0.53\%$$

10. Presentation of errors

Error in a number can be expressed in three different forms: (i) Absolute error, (ii) Relative error, and (iii) Percentage error.

10.1 Absolute error

The absolute error in a measured/calculated value of a physical quantity is the difference between the measured/calculated value and its true value. If N denotes the measured/calculated value of a physical quantity, then the absolute error is ΔN. The minimum and maximum possible values of the absolute error are often referred to as the 'limiting errors'. When the measured or calculated value is expressed up to nth place, the maximum possible absolute error is half a unit in the nth place, which can be expressed as

$$\Delta N = \frac{1}{2}\left(10^{-n}\right)$$

In general, the places up to which a value is reported is decided with the help of the least count. Absolute error has the same dimensions as the physical quantity. Some examples of maximum absolute error are shown in the following table.

N	Max. value of ΔN
4.37	0.005
-0.0056	0.00005
2.67×10^6	0.005×10^6
4.236×10^{-5}	0.0005×10^{-5}
-5.47×10^9	0.005×10^9

Exercise 11: Write down the maximum value of absolute error in the numbers 4.32, 87432, 3500, 2.83×10^{-3}, 2.15×10^{3}, -3.245×10^{-2}, -3.65×10^{4}.

Solution: The given number N and maximum value of absolute error ΔN in the number are given in the following table.

N	Max. value of ΔN
4.32	0.005
87432	0.5
3500	0.5
2.83×10^{-3}	0.005×10^{-3}
2.15×10^{3}	0.005×10^{3}
-3.245×10^{-2}	0.0005×10^{-2}
-3.65×10^{4}	0.005×10^{4}

10.2　Relative error

The relative error for a physical quantity N is expressed as the ratio of the absolute error ΔN and the value N of the quantity. Hence, the relative error E_r is expressed as

$$E_r = \frac{\Delta N}{N}$$

Notice that the relative error is a dimensionless quantity.

10.3　Percentage error

The percentage error E_p for a physical quantity N is expressed as

$$E_p = \frac{\Delta N}{N} \times 100$$

Notice that the percentage error also is a dimensionless quantity.

Exercise 12. The length of a block measured with the help of vernier calliper is 5.36 cm. Calculate (i) absolute error, (ii) relative error and (iii) percentage error in the measured value.

Solution: The least count of the vernier calliper is 0.01 cm

(i) The absolute error is

$$\Delta N = \frac{0.01}{2} = 0.005 \text{ cm}$$

(ii) The relative error is

$$E_r = \frac{\Delta N}{N} = \frac{0.005}{5.36} = 0.0009$$

(iii) The percentage error is

$$E_p = \frac{\Delta N}{N} \times 100 = \frac{0.005}{5.36} \times 100 = 0.09\%$$

11. Index of accuracy

A belief about the accuracy of a result may be that it is represented by the number of decimal points in its reported value.[4] This statement cannot be correct in general. The true index of accuracy of a result is the relative error. It can be understood, for example, with the help of the following exercises.

Exercise 13: In a measurement, diameter of a 4 cm steel shaft is measured to the nearest thousandth of a centimeter. In another measurement, 2 kilometer of railroad track is measured to the nearest cm. Which of the measurements is more accurate.

Solution: Let us denote the two measurements by A and B, respectively. The absolute error in the first measurement is $\Delta A = 0.001/2 = 0.0005$ c m whereas the absolute error in the second measurement is $\Delta B = 1/2 = 0.5$ cm. According the widespread belief, the measurement A is more accurate as compared to the measurement B. But, it is not the case, as the true index of accuracy is the relative error. The relative errors in the two cases are

$$E_{ra} = \frac{\Delta A}{A} = \frac{0.0005}{4} = \frac{1}{8000}$$

and

$$E_{rb} = \frac{\Delta B}{B} = \frac{0.5}{2 \times 10^5} = \frac{1}{400000}$$

Hence, in the measurement A, there is an error of 1 part in 8000 whereas in the measurement B, there is an error of 1 part in 400000. The latter measurement is clearly more accurate.

---

[4]It is applicable when the same physical quantity is measure by two instruments having different least counts. For example, length of a small block is measured with the help of Vernier calliper and of screw gauge.

Exercise 14: In one measurement, a 4 m rod is measured to the nearest thousandth of a centimeter. In another measurement, 4 kilometer of railroad track is measured to the nearest 1 m. Which of the measurements is more accurate.

Solution: Let us denote the two measurements by A and B, respectively. The absolute error in the first measurement is $\Delta A = 0.001/2 = 0.0005$ cm whereas the absolute error in the second measurement is $\Delta B = 1/2 = 0.5$ m. In order to decide about the true accuracy, let us calculate the relative errors. The relative errors in the two cases are

$$E_{ra} = \frac{\Delta A}{A} = \frac{0.0005}{4 \times 10^2} = \frac{1}{800000}$$

and

$$E_{rb} = \frac{\Delta B}{B} = \frac{0.5}{4 \times 10^3} = \frac{1}{8000}$$

Hence, in the measurement A, there is an error of 1 part in 800000 whereas in the measurement B, there is an error of 1 part in 8000. The former measurement is clearly more accurate.

12. Error formulas

Derived quantities are obtained in terms of fundamental quantities and other derived quantities. The final result of an experiment is calculated from a number of observations taken with the help of different instruments and are connected through a formula. Here, we shall discuss error formulas for various operations.

12.1 Addition

Let us have various physical quantities u_1, u_2, ..., u_n, having the same dimensions. The addition of them is

$$N = u_1 + u_2 + \ldots + u_n$$

Then, we have

$$\Delta N = \Delta u_1 + \Delta u_2 + \ldots + \Delta u_n$$

The maximum probable errors Δu_1, Δu_2, ..., Δu_n are positive. The absolute error of a sum of approximate numbers is therefore equal to the sum of their absolute errors, .

Exercise 15: With the help of a meter-scale, length and breadth of a rectangular block are measured as 5.2 cm and 3.4 cm, respectively.

Calculate the circumference of the block and maximum probable error in the result.

Solution: We have length $l = 5.2$ cm and breadth $b = 3.4$ cm. The circumference c of the block is

$$c = 2(l + b) = 2(5.2 + 3.4) = 17.2 \text{ cm}$$

The least count of meter-scale is 0.1 cm. Thus, the maximum probable errors in the measurement of length and breadth are

$$\Delta l = 0.1/2 = 0.05 \text{ cm} \qquad \text{and} \qquad \Delta b = 0.1/2 = 0.05 \text{ cm}$$

The maximum probable error in the circumference c is

$$\Delta c = 2(\Delta l + \Delta b) = 2(0.05 + 0.05) = 0.2 \text{ cm}$$

The maximum probable error is at the first decimal place. Therefore, the result will be reported up to first decimal place along with the maximum probable error. Hence, the circumference of the block is

$$c = 17.2 \pm 0.2 \text{ cm}$$

Exercise 16: With the help of a vernier calliper, three sides of a triangle are measured as 3.25 cm, 4.32 cm, 5.12 cm. Calculate the circumference of the triangle and maximum probable error in the result.

Solution: We have three sides of a triangle $a = 3.25$ cm, $b = 4.32$ cm, $c = 5.12$ cm. The circumference s of the triangle is

$$s = a + b + c = 3.25 + 4.32 + 5.12 = 12.69 \text{ cm}$$

The least count of vernier calliper is 0.01 cm. Thus, the maximum probable errors in the measurement of sides of the triangle are

$$\Delta a = 0.01/2 = 0.005 \text{ cm} \qquad\qquad \Delta b = 0.01/2 = 0.005 \text{ cm}$$

$$\Delta c = 0.01/2 = 0.005$$

The maximum probable error in the circumference c is

$$\Delta s = \Delta a + \Delta b + \Delta c = 0.005 + 0.005 + 0.005 = 0.015 \text{ cm}$$

The maximum probable error is at the second decimal place. Therefore, the result will be reported up to second decimal place along with the maximum probable error. Hence, the circumference is

$$c = 12.69 \pm 0.02 \text{ cm}$$

12.2 Subtraction

Let us have two physical quantities u_1 and u_2 having the same dimensions. The subtraction of u_2 from u_1 is

$$N = u_1 - u_2$$

Then we have

$$\Delta N = \Delta u_1 - \Delta u_2$$

The maximum probable errors Δu_1 and Δu_2 are positive. We must take the sum of the absolute values of errors in order to get the maximum possible error in N. Thus, we have

$$\Delta N = \Delta u_1 + \Delta u_2$$

Remark: Notice that this expression is the same as would have been for the addition of two numbers u_1 and u_2.

Exercise 17: Length of two rods measured with the help of a meter-scale are 10.2 cm and 9.5 cm. Calculate the difference between the lengths of two rods and maximum probable error in the result.

Solution: Length of two the rods are $l_1 = 10.2$ cm and $l_2 = 9.5$ cm. The difference between the lengths is

$$N = |l_1 - l_2| = |10.2 - 9.5| = 0.7 \text{ cm}$$

The least count of meter-scale is 0.1 cm. Thus, the maximum probable errors in the measurement of lengths are

$$\Delta l_1 = 0.1/2 = 0.05 \text{ cm} \qquad \text{and} \qquad \Delta l_2 = 0.1/2 = 0.05 \text{ cm}$$

The maximum probable error in the difference is

$$\Delta N = \Delta l_1 + \Delta l_2 = 0.05 + 0.05 = 0.1 \text{ cm}$$

The maximum probable error is at the first decimal place. Therefore, the result will be reported up to first decimal place along with the maximum probable error. Hence, the difference between the two lengths is

$$N = 0.7 \pm 0.1 \text{ cm}$$

Exercise 18: Diameters of two wires measured with the help of a screw gauge are 4.36mm and 2.89 mm. Calculate the difference between the radii of two wires and maximum probable error in the result.

Solution: Diameters of two wires are $d_1 = 4.36$ mm and $d_2 = 2.89$ mm. The difference between the radii is

$$N = |d_1 - d_2| = |4.36 - 2.89| = 1.47 \text{ mm}$$

The least count of screw gauge is 0.01 mm. Thus, the maximum probable errors in the measurement of diameters are

$$\Delta d_1 = 0.01/2 = 0.005 \text{ mm} \qquad \text{and} \qquad \Delta d_2 = 0.01/2 = 0.005 \text{ mm}$$

The maximum probable error in the difference is

$$\Delta N = \Delta d_1 + \Delta d_2 = 0.005 + 0.005 = 0.01 \text{ mm}$$

The maximum probable error is at the second decimal place. Therefore, the result will be reported up to second decimal place along with the maximum probable error. Hence, the difference between the two diameters is

$$N = 1.47 \pm 0.01 \text{ mm}$$

12.3 Multiplication

Let us have various physical quantities u_1, u_2, ..., u_n (not necessarily having the same dimensions). The multiplication of them is

$$N = u_1 \, u_2 \, \ldots \, u_n$$

In order to calculate the maximum probable error in N, we proceed in the following manner. On taking logarithm, we have

$$\ln N = \ln u_1 + \ln u_2 + \ldots + \ln u_n$$

On differentiation, we get

$$\frac{\Delta N}{N} = \frac{\Delta u_1}{u_1} + \frac{\Delta u_2}{u_2} + \ldots + \frac{\Delta u_n}{u_n}$$

The maximum probable errors Δu_1, Δu_2, ..., Δu_n are positive. The maximum probable error in the product N is

$$\Delta N = N \left[\frac{\Delta u_1}{u_1} + \frac{\Delta u_2}{u_2} + \ldots + \frac{\Delta u_n}{u_n} \right]$$

Exercise 19: With the help of a meter-scale, length and breadth of a rectangular block are measured as 4.2 cm and 2.9 cm, respectively.

Calculate the area of the block and maximum probable error in the result.

Solution: We have length $l = 4.2$ cm and breadth $b = 2.9$ cm. The area a of the block is

$$a = l \times b = 4.2 \times 2.9 = 12.18 \text{ cm}^2$$

The least count of meter-scale is 0.1 cm. Thus, the maximum probable errors in the measurement of length and breadth are

$$\Delta l = 0.1/2 = 0.05 \text{ cm} \qquad \text{and} \qquad \Delta b = 0.1/2 = 0.05 \text{ cm}$$

The maximum probable error in the area a is

$$\Delta a = a\left[\frac{\Delta l}{l} + \frac{\Delta b}{b}\right] = 12.18\left[\frac{0.05}{4.2} + \frac{0.05}{2.9}\right] = 0.355 \text{ cm}^2$$

The maximum probable error is at the first decimal place. Therefore, the result will be reported up to first decimal place along with the maximum probable error. Hence, the area is

$$a = 12.2 \pm 0.4 \text{ cm}^2$$

12.4 Division

Let us have two physical quantities u_1 and u_2 (not necessarily having the same dimensions). The division 0f u_1 by u_2 is

$$N = \frac{u_1}{u_2}$$

In order to calculate the maximum probable error in N, we proceed in the following manner. On taking logarithm, we have

$$\ln N = \ln u_1 - \ln u_2$$

On differentiating it, we get

$$\frac{\Delta N}{N} = \frac{\Delta u_1}{u_1} - \frac{\Delta u_2}{u_2}$$

The maximum probable errors Δu_1 and Δu_2 are positive. We must take them with the positive sign in order to be sure of the maximum error in the function N. The maximum probable error in the division N is

$$\Delta N = N\left[\frac{\Delta u_1}{u_1} + \frac{\Delta u_2}{u_2}\right]$$

Remark: Notice that this expression is the same as would have been for the multiplication of two numbers u_1 and u_2.

Exercise 20: A solid body has mass 10.9 gm and volume 5.23 cm^3. Calculate its density and maximum probable error in the result.

Solution: We have mass $m = 10.9$ gm and volume $v = 5.23$ cm^3. The density d of the body is

$$d = \frac{m}{v} = \frac{10.9}{5.23} = 2.08413 \text{ gm/m}^2$$

The maximum probable errors in the measurement of lengths are

$$\Delta m = 0.1/2 = 0.05 \text{ cm} \qquad \text{and} \qquad \Delta v = 0.01/2 = 0.005 \text{ cm}$$

The maximum probable error in the density d is

$$\Delta d = d\left[\frac{\Delta m}{m} + \frac{\Delta v}{v}\right] = 2.08413\left[\frac{0.05}{10.9} + \frac{0.005}{5.23}\right] = 0.0116 \text{ gm/cm}^3$$

The maximum probable error is at the second decimal place. Therefore, the result will be reported up to second decimal place along with the maximum probable error. Hence, the density is
$$d = 2.08 \pm 0.01 \text{ gm/cm}^3$$

12.5 Powers and roots

For a physical quantity N, let us have

$$N = ku^m$$

where k and m are constants. When m is a positive integer, it is the case of a power, and when m is $1/r$, where r is a positive integer, it is the case of a root. In general, m may be any positive or negative real number. For the maximum relative error, we first take logarithm, so that

$$\ln N = \ln k + m \ln u$$

On differentiation, we get

$$\frac{\Delta N}{N} = m \frac{\Delta u}{u}$$

Thus, the maximum probable error in N is

$$\Delta N = mN \frac{\Delta u}{u}$$

Exercise 21: By using screw gauge, diameter of a wire is measured as 2.45 mm. Calculate the area of cross section of the wire and maximum probable error in the result.

Solution: We have diameter $d = 2.45$ mm. The area of cross section a of the wire is

$$a = \frac{\pi d^2}{4} = \frac{3.14 \times (2.45)^2}{4} = 4.71196 \text{ mm}^2$$

The least count of screw gauge is 0.01 mm. Thus, the maximum probable error in the measurement of diameter is

$$\Delta d = 0.01/2 = 0.005 \text{ mm}$$

The maximum probable error in the area a is

$$\Delta a = 2a\,\frac{\Delta d}{d} = 2 \times 4.71196 \times \frac{0.005}{2.45} = 0.019 \text{ mm}^2$$

The maximum probable error is at the second decimal place. Therefore, the result will be reported up to the second decimal place along with the maximum probable error. Hence, the area of cross-section of the wire is

$$a = 4.71 \pm 0.02 \text{ mm}^2$$

12.6 General expression

Let us consider a quantity N expressed as

$$N = \frac{K a^m b^n c^p}{d^q f^r}$$

where K is some constant, m, n, p, q, and r are powers and roots, and a, b, c, d, and f are measured or calculated values of parameters. The maximum probable error in N can be obtained in the following manner. On taking logarithm, we have

$$\ln N = \ln K + m \ln a + n \ln b + p \ln c - q \ln d - r \ln f$$

On differentiating, we get

$$\frac{\Delta N}{N} = m\,\frac{\Delta a}{a} + n\,\frac{\Delta b}{b} + p\,\frac{\Delta c}{c} - q\,\frac{\Delta d}{d} - r\,\frac{\Delta f}{f}$$

Now, m, n, p, q, r, Δa, Δb, Δc, ..., Δe are positive. We must take all the terms with the positive sign in order to get the maximum error in the expression N. Hence, the maximum probable error in N is

$$\Delta N = N\left[m\,\frac{\Delta a}{a} + n\,\frac{\Delta b}{b} + p\,\frac{\Delta c}{c} + q\,\frac{\Delta d}{d} + r\,\frac{\Delta f}{f}\right]$$

Exercise 22: Calculate the quantity N expressed as

$$N = \frac{36.5^2 \times 47.65 \times 48.35^{1/3}}{45.83^2 \times 83.2^{1/2}}$$

Solution: The calculated value of N is 12.07126. The expression for relative error is

$$\frac{\Delta N}{N} = 2\,\frac{0.05}{36.5} + \frac{0.005}{47.65} + \frac{1}{3}\,\frac{0.005}{48.35} + 2\,\frac{0.005}{45.83} + \frac{1}{2}\,\frac{0.05}{83.2} = 3.398 \times 10^{-3}$$

Hence,

$$\Delta N = (3.398 \times 10^{-3}) \times (12.07126) = 0.041$$

As the error is at the second decimal place, we have $N = 12.07 \pm 0.04$.

Exercise 23: For the quantity N expressed as

$$N = \frac{K(p+q)^m\, s^p}{(w-x-y)^q (z+u-v)^r}$$

where K is some constant, m, p, q, and r are powers and roots, and p, q, s, w, x, y, z, u and v are measured or calculated values of parameters. Obtain an expression for the maximum relative error in N.

Solution: Let us first express

$$a = p + q \qquad\qquad b = w - x - y \qquad\qquad c = z + u - v$$

so that

$$N = \frac{K a^m\, s^p}{b^q\, c^r}$$

On taking logarithm, we have

$$\ln N = \ln K + m \ln a + p \ln s - q \ln b - r \ln c$$

On differentiating, we get

$$\frac{\Delta N}{N} = m\,\frac{\Delta a}{a} + p\,\frac{\Delta s}{s} - q\,\frac{\Delta b}{b} - r\,\frac{\Delta c}{c}$$

Now, m, p, q, r, Δa, Δs, Δb, Δc are positive. We must take all the terms with the positive sign in order to get the maximum probable error in the expression N. Hence, we have

$$\Delta N = N\left[m\,\frac{\Delta a}{a} + p\,\frac{\Delta s}{s} + q\,\frac{\Delta b}{b} + r\,\frac{\Delta c}{c} \right]$$

Here, we have

$$\Delta a = \Delta p + \Delta q \qquad \Delta b = \Delta w + \Delta x + \Delta y \qquad \Delta c = \Delta z + \Delta u + \Delta v$$

Exercise 24: Calculate the quantity N expressed as

$$N = \frac{(15.3 + 20.12)^2 \times (73.85 - 33.2)^3}{(22.2 - 2.56)^2 \times (55.2 + 32.56 - 10.2)^{1/2}}$$

Solution: The calculated value of N is 2.4807×10^4. The relative error is

$$\frac{\Delta N}{N} = 2\,\frac{0.05 + 0.005}{35.42} + 3\,\frac{0.005 + 0.05}{40.65} + 2\,\frac{0.05 + 0.005}{19.64}$$

$$+ \frac{1}{2}\,\frac{0.05 + 0.005 + 0.05}{77.56} = 0.01344$$

Therefore,

$$\Delta N = 2.4807 \times 10^4 \times 0.01344 = 0.033 \times 10^4$$

For the data expressed in 10^4, the error is at the second decimal place, Thus, we have $N = (2.48 \pm 0.03) \times 10^4$.

13. Elements of statistics

Here, we shall discuss about some basic methods used in statistics. We are aware of the fact that statistics is applied when the data set is quite large.

13.1 Arithmetic mean

Arithmetic mean (often called, mean) of data x_i, $i = 1, 2, \ldots, n$ is expressed as

$$\overline{x} = \frac{x_1 + x_2 + \ldots + x_n}{n} = \frac{\sum x_i}{n}$$

Exercise 25: Calculate the mean of 15, 19, 23, 35, 46.

Solution: There are five data, therefore $n = 5$. The mean (or arithmetic mean) is

$$\overline{x} = \frac{\sum x_i}{n} = \frac{15 + 19 + 23 + 35 + 46}{5} = \frac{138}{5} = 27.6$$

13.1.1 When the data have various frequencies

There may be a situation that some or all values of the data have more than one frequency. Suppose, a value x_i has a frequency f_i. Then, the arithmetic mean $\overline{x}$ of the data is as the following.

$$\overline{x} = \frac{\sum f_i x_i}{\sum f_i}$$

Exercise 26: Calculate the arithmetic mean of the data whose values x_i and their frequencies f_i are as the following.

$x_i =$	16	20	34	47	63	56	39	33	64	49
$f_i =$	3	4	3	6	3	6	4	3	6	5

Solution: The arithmetic mean of the given data can be calculated in the following manner.

x_i	f_i	$f_i x_i$
16	3	48
20	4	80
34	3	102
47	6	282
63	3	189
56	6	336
39	4	156
33	3	99
64	6	384
49	5	245
	$\sum f_i = 43$	$\sum f_i x_i = 1921$

The arithmetic mean M of the given data is

$$\overline{x} = \frac{1921}{43} = 44.67$$

13.1.2 When the data are given in the intervals

When the data are given in various intervals, the mid-value of an interval is taken as the value x_i of data. Then, the arithmetic mean $\overline{x}$ of the data is as the following.

$$\overline{x} = \frac{\sum f_i x_i}{\sum f_i}$$

Exercise 27: Calculate the mean of the data whose intervals and their frequencies are as the following.

int =	10– 20	20 – 30	30 – 40	40 – 50	50 – 60	60 – 70
$f_i =$	3	4	2	5	5	3

Solution: The arithmetic mean of the given data can be calculated in the following manner.

Range	x_i	f_i	$f_i x_i$
10 – 20	15	3	45
20 – 30	25	4	100
30 – 40	35	2	70
40 – 50	45	5	225
50 – 60	55	5	275
60 – 70	65	3	195
		$\sum f_i = 22$	$\sum f_i x_i = 910$

The arithmetic mean $\bar{x}$ of the given data is

$$\bar{x} = \frac{910}{22} = 41.36$$

13.2 Median

When we arrange data in a sequence (increasing or decreasing), the middle value in the data set is known as the median of data. When there are two values in the middle, average of them is the median of data.

Remark: For a symmetrical distribution of data, the mean and median of the data set are equal.

Exercise 28: Calculate the median of 55,20,22,35,47,82,65.

Solution: Let us first arrange the data in a sequence, say, in the increasing sequence as:

$$20, 22, 35, 47, 55, 65, 82$$

There are seven data and thus the fourth data is in the middle. Hence, the median of data is 47.

Exercise 29: Calculate the median of 54, 15, 30, 45, 82, 65, 34, 27.

Solution: Let us first arrange the data in a sequence, say, in the increasing sequence as:

$$15, \ 27, \ 30, \ 34, \ 45, \ 54, \ 65, \ 82$$

There are eight data and thus fourth and fifth data, 34 and 45, respectively, are in the middle. Hence, the median of data is

$$\text{Median} = \frac{34 + 45}{2} = 39.5$$

13.2.1 When the data have various frequencies

When the data x_i are arranged in an ascending or descending order along with their frequencies f_i, the median can be calculated in the manner as discussed in the following exercise.

Exercise 30: Calculate the median of the data whose values x_i and their frequencies f_i are as the following.

$x_i =$	15	82	23	65	35	72	47	33	19	55
$f_i =$	3	4	6	3	5	4	6	3	2	5

Solution: First the data are arranged in the descending or ascending order, and a table of their values, frequencies and cumulative frequencies are prepared in the following manner. Here, the data are arranged in the ascending order.

x_i	f_i	cf
15	3	3
19	2	5
23	6	11
33	3	14
35	5	19
47	6	25
55	5	30
65	3	33
72	4	37
82	4	41

Here, cf denotes the cumulative frequency. Total frequency (*i.e.*, total number of data) is 41. The middle data is the twenty first. The cumulative frequency 21 falls on the value 47. Thus, the median of the data is 47.

When the total number of data is an even number, there will be two data in the middle. When the cumulative frequencies corresponding to these data fall on a single number, then that number is the median. When the cumulative frequencies corresponding to these data fall on two different numbers, then the average of these numbers is the median, as discussed in the following exercise.

Exercise 31: Calculate the median of the data whose values x_i and their frequencies f_i are as the following.

$x_i =$	15	69	79	19	26	37	35	47	52	65
$f_i =$	3	7	4	2	5	4	4	6	5	3

Solution: Here, data are arranged in the ascending order. A table of the values of data, their frequencies and cumulative frequencies is prepared in the following manner.

x_i	f_i	cf
15	3	3
19	2	5
26	5	10
35	4	14
37	4	18
47	6	24
52	5	29
65	3	32
69	7	39
79	4	43

Total frequency (*i.e.*, total number of data) is 43. The middle data is the twenty second. The cumulative frequency 22 falls on the value 47. Thus, the median of the data is 47.

13.2.2 When the data are given in the intervals

When data are arranged in the intervals, then the median Me of the data is

$$Me = L_1 + \frac{L_2 - L_1}{f} \left(\frac{m}{2} - C\right)$$

where L_1, L_2 and f are the lower limit, upper limit and frequency of the median interval, respectively, m the sum of all the frequencies, and C the cumulative frequency of the interval just before the median interval.

Exercise 32: Calculate the median of the data whose intervals and their frequencies are as the following.

int $=$	$10 - 20$	$20 - 30$	$30 - 40$	$40 - 50$	$50 - 60$	$60 - 70$
$f_i =$	4	5	3	2	5	4

Solution: Here, data are arranged in the ascending order and there is no need to arrange them. A table of the values of intervals, their frequencies and cumulative frequencies are prepared in the following manner.

Range	f_i	cf
10 – 20	4	4
20 – 30	5	9
30 – 40	3	12
40 – 50	2	14
50 – 60	5	19
60 – 70	4	23

There are 23 data and therefore, the 12th data falls in the range 30 – 40. Hence, the median exists in the interval 30 – 40. Thus, we have $L_1 = 30$, $L_2 = 40$, $m = 23$, $C = 9$, $f = 3$. Hence, the median is

$$Me = L_1 + \frac{L_2 - L_1}{f} \left(\frac{m}{2} - C \right)$$

$$= 30 + \frac{40 - 30}{3} \left(\frac{23}{2} - 9 \right) = 38.33$$

13.3 Mode

In the given set of data, the value having the largest frequency is known as the mode of the data.

Exercise 33: Calculate the mode of the data whose values x_i and their frequencies f_i are as the following.

$x_i =$	38	21	24	33	44	36	47	62	70	86
$f_i =$	3	4	6	2	4	8	6	3	2	5

Solution: In the given data, the value 36 has the largest frequency 8, hence the mode of the data is 36.

13.3.1 When the data are given in the intervals

When the data are arranged in the intervals, mode Mo is as the following.

$$Mo = L_1 + \frac{f - f_1}{2f - f_1 - f_2} (L_2 - L_1)$$

where L_1 and L_2 are the lower and upper limits, respectively, of the mode interval, f, f_1 and f_2 are frequencies of the mode interval, the interval just before the mode interval and the the interval just after the mode interval, respectively.

Exercise 34: Calculate the mode of the data whose intervals and their frequencies are as the following.

int =	$10 - 20$	$20 - 30$	$30 - 40$	$40 - 50$	$50 - 60$	$60 - 70$
$f_i =$	4	6	5	7	2	3

Solution: Here, data are arranged in the ascending order and there is no need to rearrange them. The largest frequency 7 is of the interval $40 - 50$ and the mode of the data exists in this interval. Thus, we have $L_1 = 40$, $L_2 = 50$, $f = 7$, $f_1 = 5$ and $f_2 = 2$. Hence, the mode is

$$Mo = L_1 + \frac{f - f_1}{2f - f_1 - f_2} (L_2 - L_1)$$

$$= 40 + \frac{7 - 5}{14 - 5 - 2} (50 - 40) = 42.86$$

13.4 Variance

When the values of data in a set lie close to their mean value, the dispersion is less as compared to the situation when the values lie far away from their mean value. This tendency of data is expressed with the help of a parameter, called the variance and denoted by s^2 and expressed as

$$s^2 = \left[\sum_{i=1}^{n} (x_i - \bar{x})^2 \right] \Big/ (n - 1)$$

Here, n is the number of values in the set. The reason for dividing by $(n - 1)$ rather than by n is the theoretical consideration referred to as the degree of freedom. From the practical point of view, dividing the squared differences by $(n - 1)$ rather than n is necessary in order to use sample variance. The population variance, denoted by σ^2, is expressed as

$$\sigma^2 = \left[\sum_{i=1}^{n} (x_i - \bar{x})^2 \right] \Big/ n$$

13.5 Standard deviation

The variance represents squared units and, therefore, is not an appropriate measure of dispersion when we wish to express this concept in terms of the original units. To obtain the measure of dispersion in original units, we merely take the square root of the variance. The result is known as the standard deviation, denoted by s

$$s = \sqrt{\left[\sum_{i=1}^{n}(x_i - \overline{x})^2\right] \Big/ (n-1)}$$

It is the standard deviation of a sample. The standard deviation of a finite population is

$$\sigma = \sqrt{\left[\sum_{i=1}^{n}(x_i - \overline{x})^2\right] \Big/ n}$$

Note: When the number n is very large, s and σ are almost equal to each other.

Exercise 35: Calculate standard deviation for the following data:

P =	1	2	3	4	5	6	7	8	9	10
Age =	78	25	38	44	52	28	23	54	62	31

Solution: The mean of the sample is

$$\overline{x} = \frac{\sum x}{n} = \frac{435}{10} = 43.5$$

For calculation of standard deviation, we prepare a table:

x_i	$x_i - \overline{x}$	$(x_i - \overline{x})^2$
78	34.50	1190.25
25	-18.50	342.25
38	-5.50	30.25
44	0.50	0.25
52	8.50	72.25
28	-15.50	240.25
23	-20.50	420.25
54	10.50	110.25
62	18.50	342.25
31	-12.50	156.25
435		2904.5

The standard deviation is

$$\sigma = \sqrt{\left[\sum_{i=1}^{n}(x_i - \overline{x})^2\right] \Big/ n} = \sqrt{\frac{2904.5}{10}} = 17.04$$

Exercise 36: Calculate standard deviation for the following data:

P =	1	2	3	4	5	6	7	8	9
Age =	64	29	26	81	31	82	41	72	61

Solution: The mean of the sample is

$$\overline{x} = \frac{\sum x}{n} = \frac{487}{9} = 54.11$$

For calculation of standard deviation, we prepare a table:

x_i	$x_i - \overline{x}$	$(x_i - \overline{x})^2$
64	9.89	97.79
29	-25.11	630.57
26	-28.11	790.23
81	26.89	723.01
31	-23.11	534.12
82	27.89	777.79
41	-13.11	171.90
72	17.89	320.01
61	6.89	47.46
487		4092.88

The standard deviation is

$$\sigma = \sqrt{\left[\sum_{i=1}^{n}(x_i - \overline{x})^2\right]\Big/ n} = \sqrt{\frac{4092.88}{9}} = 21.32$$

14. Multiple choice questions

1. Which of the followings is not a derived quantity.

   (i) Velocity   (ii) Acceleration   (iii) Density   (iv) Mass

   Ans. (iv)

2. Which of the followings is a derived quantity.

   (i) Velocity   (ii) Length   (iii) Time   (iv) Mass

   Ans. (i)

3. After making observations, it was found that the stop-watch was defective. This type of error is known as

   (i) Random error   (ii) Systematic error

   (iii) Truncation error   (iv) Rounding-off error

   Ans. (ii)

4. Which of the following errors cannot be rectified.

    (i) Random error (ii) Systematic error

    (iii) Truncation error (iv) Rounding-off error

    Ans. (ii)

5. The relation between the maximum uncertainty U and the least count L for a measurement is

    (i) $U = L$ (ii) $U = L/2$ (iii) $U = 2L$ (i) $U = L^2$

    Ans. (ii)

6. The meter-scale, vernier calliper, screw gauge have least counts 1 mm, 0.1 mm, 0.01 mm, respectively. Two small objects A and B have lengths 2.456 mm and 2.547 mm, respectively when measured with the help of some very accurate instrument. The lengths of the objects A and B can be resolved with the help of

    (i) meter-scale (ii) both vernier calliper and screw gauge

    (iii) vernier calliper only (iv) screw gauge only

    Ans. (ii)

7. In the number 0.00608, the number of significant digits is:

    (i) 2 (ii) 3 (iii) 5 (iv) 6

    Ans. (ii)

8. In the number 60800, the number of significant digits is:

    (i) 2 (ii) 3 (iii) 4 (iv) 5

    Ans. (iv)

9. In the number 2.5060×10^{-2}, the number of significant digits is:

    (i) 2 (ii) 3 (iii) 4 (iv) 5

    Ans. (iv)

10. In the number 20.05060, the number of significant digits is:

    (i) 2 (ii) 3 (iii) 5 (iv) 7

    Ans. (iv)

11. Least count of vernier calliper is 0.01 cm. The maximum probable error in the measurement made with the help of vernier calliper is

    (i) 0.01 cm (ii) 0.02 cm (iii) 0.005 cm (iv) 0.1 cm

    Ans. (iii)

12. Least count of screw gauge is 0.01 mm. The maximum probable error in the measurement made with the help of screw gauge is

    (i) 0.01 mm   (ii) 0.02 mm   (iii) 0.005 mm   (iv) 0.1 mm

    Ans. (iii)

13. Arithmetic mean of 9, 10, 14 and 17 is

    (i) 9   (ii) 10   (iii) 14   (iv) 17

    Ans. (ii)

14. Median of 17, 10, 8, 13, 22 is

    (i) 8   (ii) 13   (iii) 14   (iv) 10

    Ans. (ii)

15. Values x_i and their frequencies f_i are as the following.

$x_i =$	10	20	30	40	50
$f_i =$	5	7	4	3	2

    The mode of the data is

    (i) 5   (ii) 7   (iii) 4   (iv) 2

    Ans. (ii)

15. Problems and questions

1. What are physical quantities. Distinguish between the fundamental and derived quantities. Write two examples of each of them.

2. Explain with reason, which of the followings are derived quantities.

    (i) Length, (ii) Height, (iii) Volume, (iv) Velocity, (v) Acceleration, (vi) Momentum, (vii) Force, (viii) Angular momentum

3. What do you understand from the least count of an equipment? Write a relation between the least count and maximum possible uncertainty in the measurement made with the help of the equipment.

4. When measured with very accurate apparatus, length of a small object is found to be 1.34567 cm. What will be the length of the object when it is measured with the help of: (i) vernier calliper and (ii) screw gauge having the least counts 0.01 cm and 0.001 cm, respectively.

5. Measurement is taken with the help of vernier calliper having least count 0.1 mm. Calculate the maximum possible uncertainty in the measurement made with the help of the vernier calliper.

6. Discuss about the significant figures. Write down rules set up for determining the number of significant figures in a given number.

7. Explain with examples about the random and systematic errors.

8. Explain with examples about the truncation and rounding-off errors.

9. After calculations of an expression, we got a value 9.583. If we want to retain the value up to first decimal place, calculate the (i) truncation error and (ii) rounding-off error.

10. What is difference between an integer and a real number? Describe the process of rounding-off the real numbers. Round-off the following numbers up to third decimal place.

    53.67298, 91.89234, 6.48995001, 19.0465, 47.1475

11. Describe the absolute error, relative error and percentage error in a measurement. Length of an object measured with the help of vernier calliper is 4.56 cm. Calculate absolute error, relative error and percentage error in the measured value.

12. Radius of a wire measured with the help of screw gauge is 8.43 mm. Calculate (i) absolute error, (ii) relative error and (iii) percentage error in the measured value.

13. Describe the index of accuracy. In a measurement, diameter of a 3 cm steel shaft is measured to the nearest thousandth of a centimeter. In another measurement, 1.5 kilometer of road is measured to the nearest centimeter. Which of the measurements is more accurate.

14. With the help of a vernier calliper, length and breadth of a rectangular block are measured as 6.34 cm and 5.04 cm, respectively. Calculate the circumference of the block and maximum probable error in the result.

15. With the help of a meter-scale, three sides of a triangle are measured as 4.6 cm, 5.7 cm, 6.3 cm. Calculate the circumference of the triangle and maximum probable error in the result.

16. Diameters of two rods measured with the help of a vernier calliper are 10.2 mm and 9.5 mm. Calculate the difference between the lengths of two rods and maximum probable error in the result.

17. With the help of a vernier calliper, length and breadth of a rectangular block are measured as 4.23 cm and 2.94 cm, respectively. Calculate the area of the block and maximum probable error in the result.

18. A solid body has mass 15.6 gm and volume 4.83 cm^3. Calculate its density and maximum probable error in the result.

19. By using screw gauge, diameter of a wire is measured as 3.67 mm. Calculate the area of cross section of the wire and maximum probable error in the result.

20 Calculate the quantity N expressed as

$$N = \frac{(25.3 + 18.12)^2 \times (53.85 - 23.2)^3}{(21.3 - 3.46)^2 \times (54.2 + 29.56 - 11.2)^{1/2}}$$

21. Describe the arithmetic mean of data. Calculate the arithmetic mean of following values.

    20, 36, 85, 63, 66, 39, 33, 46, 75, 49, 35, 72

22. Describe the arithmetic mean when data have various frequencies. Calculate the arithmetic mean of the data whose values x_i and their frequencies f_i are as the following.

$x_i =$	19	23	34	45	62	56	38	33
$f_i =$	4	5	3	6	4	3	2	5

23. Describe the arithmetic mean when data are expressed in various intervals. Calculate the mean of the data whose intervals and their frequencies are as the following.

int $=$	10– 20	20 − 30	30 − 40	40 − 50	50 − 60
$f_i =$	3	2	4	5	2

24. Describe the median of data. Calculate the median of following data.

    15, 20, 35, 87, 67, 56, 38, 33, 47.

25. Describe the median of data. Calculate the median of following data.

    15, 20, 35, 87, 67, 56, 38, 33, 47, 54.

26. Describe the median when data have various frequencies. Describe the median of Calculate the median of the data whose values x_i and their frequencies f_i are as the following.

$x_i =$	16	79	28	62	42	72	47	33
$f_i =$	4	5	3	3	5	4	2	5

27. Describe the median when data are expressed in various intervals. Calculate the median of the data whose intervals and their frequencies are as the following.

int $=$	$10 - 20$	$20 - 30$	$30 - 40$	$40 - 50$	$50 - 60$
$f_i =$	4	3	5	3	7

28. Describe the mode of data. Calculate the mode of the data whose values x_i and their frequencies f_i are as the following.

$x_i =$	45	21	24	33	54	38	47	64
$f_i =$	3	4	2	4	8	6	3	2

29. Describe the mode when data are expressed in various intervals. Calculate the mode of the data whose intervals and their frequencies are as the following.

int $=$	$10 - 20$	$20 - 30$	$30 - 40$	$40 - 50$	$50 - 60$
$f_i =$	3	4	5	7	4

30. Write notes on the following

    (i) Fundamental quantities

    (ii) Derived quantities

    (iii) Least count of an instrument

    (iv) Random and systematic error

    (v) Truncation and rounding-off error

    (vi) Absolute, relative and percentage errors

    (vii) Index of accuracy

II. Optics

The electromagnetic spectrum consists of gamma-rays, x-rays, ultraviolet radiation, visible light, infrared radiation, micro waves and radio waves. Our eyes can see only visible light, which has the radiations having wavelengths from 4000 Å to 7500 Å. Optics is the science which deals with the visible light. In this chapter, we shall discuss about the optics.

1. Lenses of various shapes

When we have a glass enclosed by two smooth surfaces, it is termed as a lens. Taking into account the shapes of the two surfaces, the lenses may be classified into six kinds, shown in Figure 1: (i) double convex lens (or simply, convex lens), (ii) plano convex lens (one of the surfaces is plane), (iii) concavoconvex lens, (iv) double concave lens (or simply, concave lens), (v) plano concave lens (one of the surfaces is plane) and (vi) convexoconcave lens. Notice that (iii) and (vi) are the same. In the present discussion, we shall mainly discuss about the convex lens and concave lens. The thickness of each lens is very small. So, we shall deal with the thin lenses.

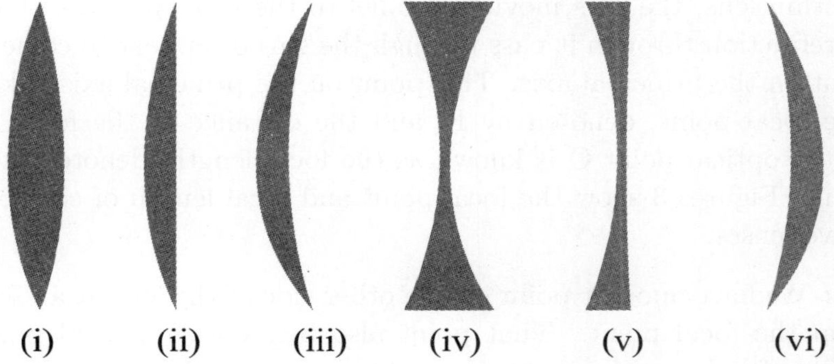

Figure 1: Lenses of various kinds.

1.1 Principal axis

The line connecting the centers of curvatures of two faces of a lens is known as the principal axis. For example for a convex lens or a concave lens shown in Figure 2, the C_1 and C_2 are centers of curvatures of two faces of the lens. The line joining the C_1 and C_2 is known as the principal axis.

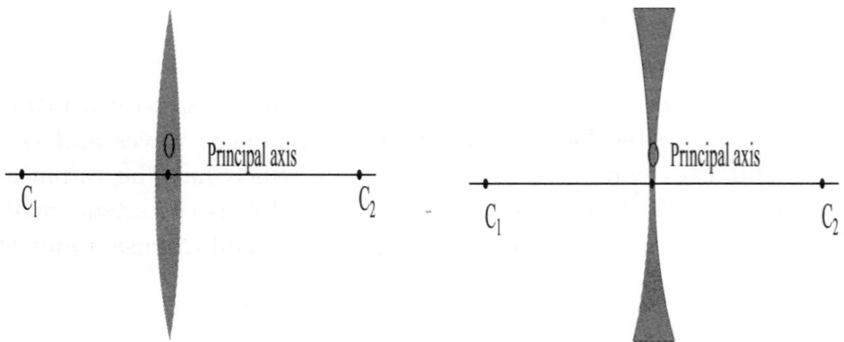

Figure 2: The centers of curvatures C_1 and C_2 for the faces of a lens are at very large distances.

1.2 Optical point

The point of intersection of the principal axis with the lens is known as the optical point. For a convex lens and a concave lens, the optical point denoted by O is shown in Figure 2.

1.3 Focal length

For a thin lens, the rays moving parallel to the principal axis of a lens after refraction through it pass through the lens or appear to come from a point on the principal axis. This point on the principal axis is known as the focal point, denoted by F, and the distance of the focal point from the optical point O is known as the focal length, denoted by f of the lens. Figures 3 show the focal point and focal length of convex and concave lenses.

Note: We have another point on the other side of the lens at a distance f from the focal point. That point also behaves as a focal point as the rays of light coming from the right (opposite) side will come to or appears to start from that point.

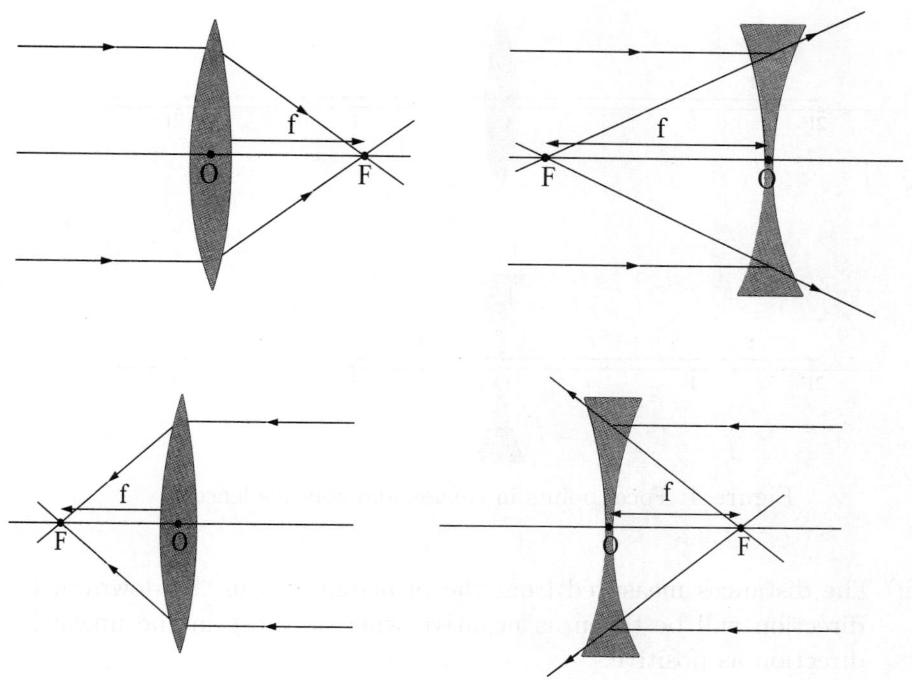

Figure 3: Focal points and focal lengths of convex and concave lenses.

1.4 Point 2F

On each side of a lens, we have one point at a distance $2f$ from the optical point. This point is generally denoted by 2F and is known as the second focal point. The F and 2F for convex and concave lenses are shown in Figure 4.

1.5 Sign convention

For formation of image of an object placed in front of a lens, we shall put the principal axis of the lens along a horizontal direction. An object will be placed on the left side to the lens. Now, for the distances measured, the sign convention is as the following (Figure 5).

(i) The distances measured from the optical point in the left direction will be taken as negative whereas those in the right direction as positive.

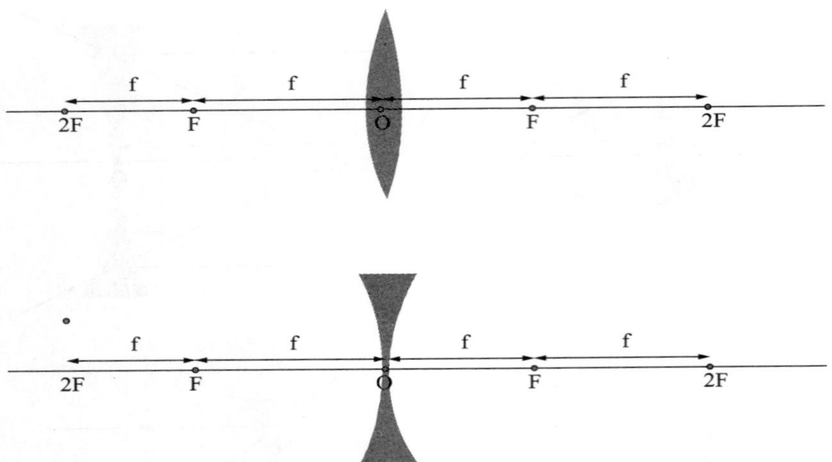

Figure 4: Focal points in convex and concave lenses.

(ii) The distances measured from the principal axis in the downward direction will be taken as negative whereas those in the upward direction as positive.

For example, in the following figure for a lens, the principal axis is CC' and the optical point O. Now, a distance measured from O in the left direction is negative $(-)$ whereas that in the right direction is positive $(+)$. Further, a distance measured from the principal axis in the upward direction is positive $(+)$ whereas that in the downward direction is negative $(-)$.

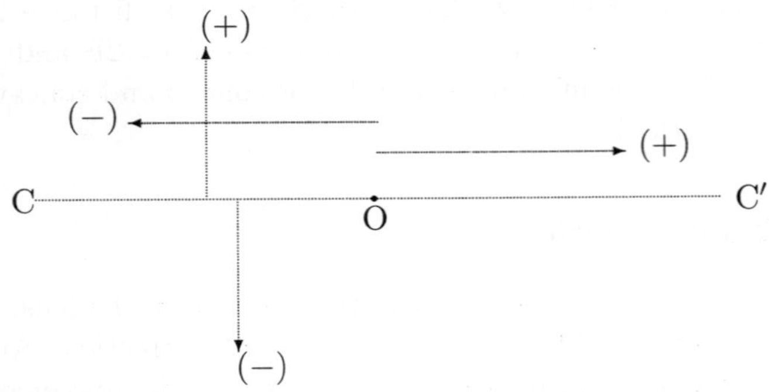

Figure 5: Sign convention.

Thus, the focal length of a convex lens is positive whereas that of a concave lens is negative.

2. Expression for focal length of a lens

For convenience, let us consider a thin convex lens of refractive index μ relative to air as shown in Figure 6. The centers of curvatures of the surfaces A and B are C_1 and C_2 and the radii of curvatures are R_1 and R_2, respectively. The focal length f of the lens is

$$\frac{1}{f} = (\mu - 1)\left(\frac{1}{R_1} - \frac{1}{R_2}\right)$$

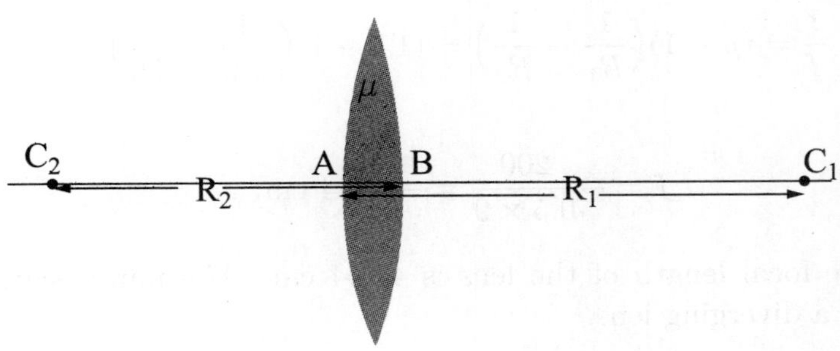

Figure 6: Convex thin lens enclosed between two spherical surfaces A and B of radii of curvatures R_1 and R_2, respectively.

Note: This expression for the focal length of a lens is valid for all shapes of lenses. However, one has to account for the sign convention for the curvatures R_1 and R_2. For example, in case of a convex lens (double convex lens), the value of R_1 is used with plus sign and that of R_2 with minus sign; in case of a concave lens (double concave lens), the value of R_1 is used with minus sign and that of R_2 with plus sign.

Exercise 1: Radii of curvatures of two faces of a convex lens placed in air are 30 cm and 40 cm. If the refractive index of the material of the lens is 1.5, calculate the focal length of the lens.

Solution: Considering the sign convention, for a convex lens, we have $R_1 = 30$ cm and $R_2 = -40$ cm. For the refractive index $\mu = 1.5$, the focal length f of the lens is

$$\frac{1}{f} = (\mu - 1)\left(\frac{1}{R_1} - \frac{1}{R_2}\right) = (1.5 - 1)\left(\frac{1}{30} - \frac{1}{-40}\right)$$

or

$$f = \frac{120}{0.5 \times 7} = 34.29 \text{ cm}$$

Thus, the focal length of the lens is 34.29 cm. The plus sign shows that it is a converging lens.

Exercise 2: Radii of curvatures of two faces of a concave lens placed in air are 40 cm and 50 cm. If the refractive index of the material of the lens is 1.5, calculate the focal length of the lens.

Solution: Considering the sign convention, for a concave lens, we have $R_1 = -40$ cm and $R_2 = 50$ cm. For the refractive index $\mu = 1.5$, the focal length f of the lens is

$$\frac{1}{f} = (\mu - 1)\left(\frac{1}{R_1} - \frac{1}{R_2}\right) = (1.5 - 1)\left(\frac{1}{-40} - \frac{1}{50}\right)$$

or

$$f = -\frac{200}{0.5 \times 9} = -44.44 \text{ cm}$$

Thus, the focal length of the lens is 44.44 cm. The minus sign shows that it is a diverging lens.

Exercise 3: If radii of curvatures of two faces of a concavoconvex (convexoconcave) lens are equal, show that the lens behaves like a plane transparent plate.

Solution: Let the radius of curvature of each surface of a concavoconvex (convexoconcave) lens is r, then accounting for the sign convention, we have $R_1 = r$ and $R_2 = r$. The focal length f of the lens of refractive μ is

$$\frac{1}{f} = (\mu - 1)\left(\frac{1}{R_1} - \frac{1}{R_2}\right) = (\mu - 1)\left(\frac{1}{r} - \frac{1}{r}\right) = 0$$

Thus, the focal length $f = \infty$. It shows that the lens behaves like a plane transparent plate having refractive index μ.

3. Refraction through a lens

Though a ray which incident on a lens is refracted at both surfaces of the lens, but for convenience in a ray-diagram in the present presentation, we shall draw the incident ray up to the center of the lens and the emergent ray from the center of the lens. Before telling about the formation of image of an object with the help of a lens, let us first understand some important aspects about a lens. For formation of image of an object, we shall consider the following rays.

(i) A ray moving parallel to the principal axis of the lens after refraction passes through the focal point F in case of a convex lens or appears to move away from the focal point F in case of a concave lens, as shown in Figure 7.

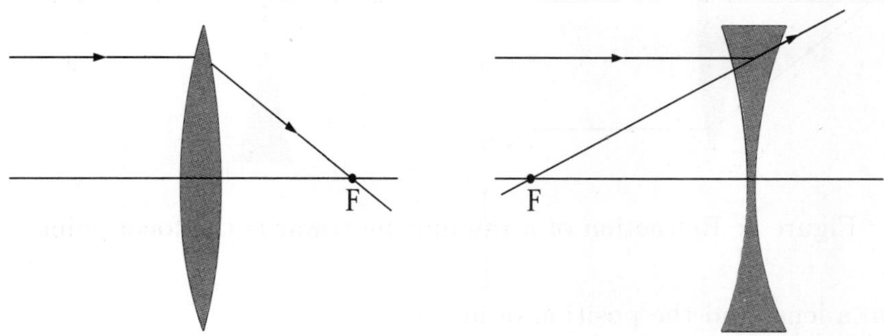

Figure 7: Refraction of a ray moving parallel to the principal axis of a lens.

(ii) A ray moving towards the optical point is refracted along the same direction. That is, this ray moves without any deviation in its path. It is the case with both the convex lens as well as the concave lens, as shown in Figure 8.

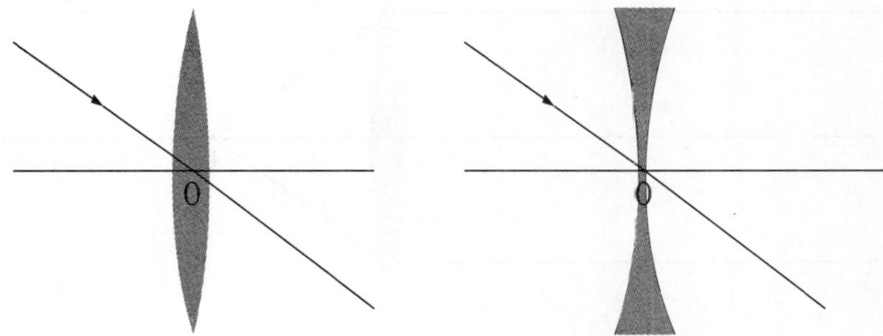

Figure 8: Refraction of a ray moving towards the optical point.

(iii) A ray moving towards the focal point after refraction becomes parallel to the principal axis of the lens, as shown in Figure 9.

4. Formation of image by a convex lens

In case of a convex lens, position of image of an object depends on the position of the object. We shall put the object at different distances

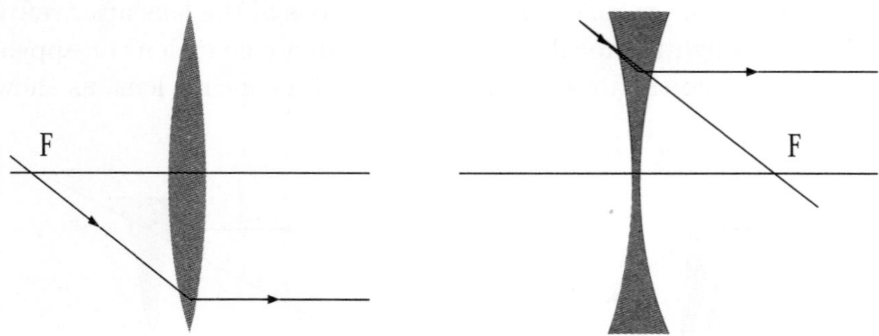

Figure 9: Refraction of a ray moving towards the focal point.

from a lens, find the position of image.

4.1 Object is at infinite distance

Suppose, an object is at infinite distance from the lens (Figure 10). The coming rays are parallel to the principal axis of the lens. These rays after refraction through the lens reach the focal point F. Thus, the image of the object is formed at the focal point of the lens.

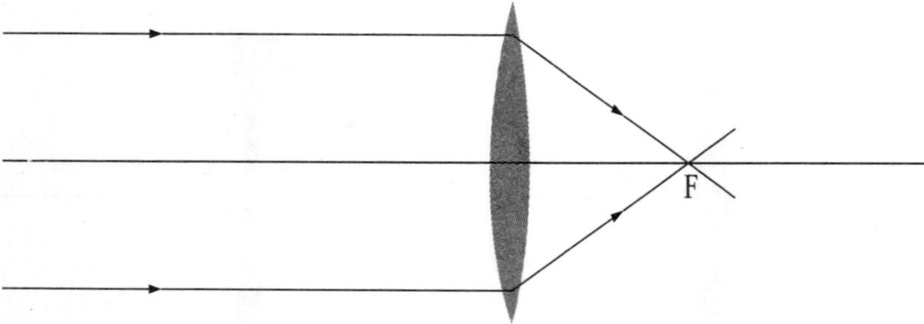

Figure 10: Formation of image when an object is placed at infinite distance from a lens.

4.2 Object is between infinity and 2F

Suppose, an object AB is placed between the infinity and 2F (Figure 11). From a point A on the object, a ray parallel to the principal axis of the lens after refraction passes through the focal point F. Other ray from the point A passing through the optical point O goes in the same direction. These two rays form A intersect at the point A'. This point

A' is the image of the point A. Similarly, we can consider other points on the object and get the corresponding images. All these images together form the image $A'B'$ of the object AB. In this case, the image is formed between the points F and 2F on the other side.

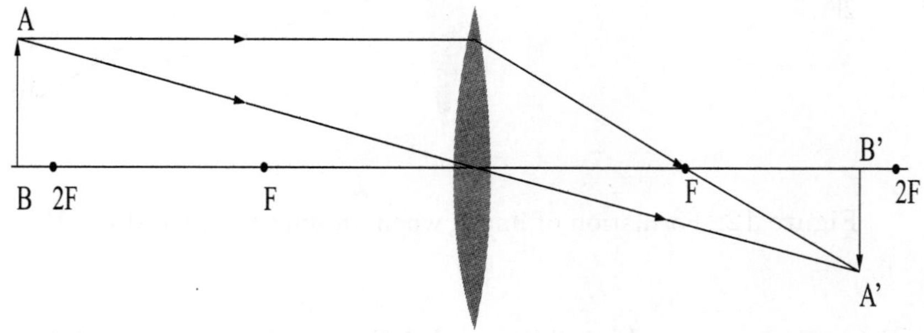

Figure 11: Formation of image when an object is placed between infinity and the point 2 F.

4.3 Object is at 2F

Suppose, an object AB is placed at 2F (Figure 12). From a point A on the object, a ray parallel to the principal axis of the lens after refraction passes through the focal point F. Other ray from the point A passing through the optical point O goes in the same direction. These two rays from A intersect at the point A'. This point A' is the image of the point A. Similarly, we can consider other points on the object and get the corresponding images. All these images together form the image $A'B'$ of the object AB. In this case, the image is formed at the point 2 F on the other side of the lens.

4.4 Object is between F and 2F

Suppose, an object AB is placed between F and 2F (Figure 13). From a point A on the object, a ray parallel to the principal axis of the lens after refraction passes through the focal point F. Other ray from the point A passing through the optical point O goes in the same direction. These two rays from A intersect at the point A'. This point A' is the image of the point A. Similarly, we can consider other points on the object and get the corresponding images. All these images together form the image $A'B'$ of the object AB. In this case, the image is formed between the point 2 F and infinity on the other side of the lens. Notice that when

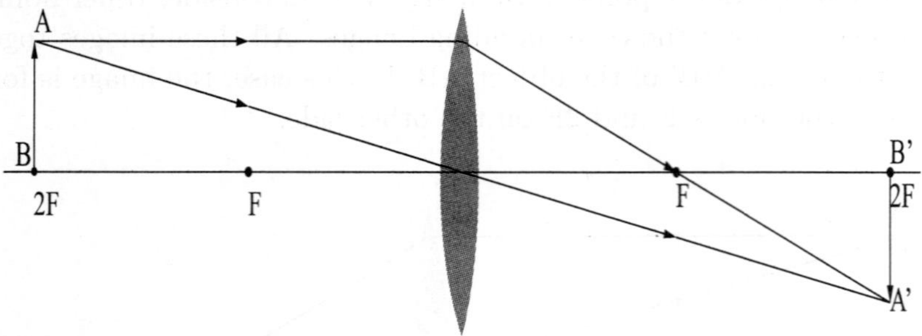

Figure 12: Formation of image when an object is placed at 2F.

an object is between the infinity and 2 F, the image is formed between the points F and 2 F and vice versa.

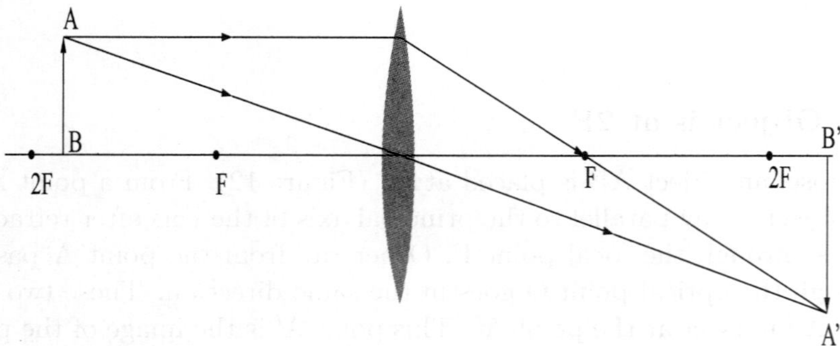

Figure 13: Formation of image when an object is placed between F and 2 F.

4.5 Object is at the focal point F

Suppose, an object AB is placed at the focal point F (Figure 14). From a point A on the object, a ray parallel to the principal axis of the lens after refraction passes through the focal point F. Other ray from the point A passing through the optical point O goes in the same direction. These two rays from A become parallel and do not intersect any where. Thus, the image of the object is formed at an infinite distance. Notice that when an object is at infinite distance, the image is formed at the focal point and vice versa.

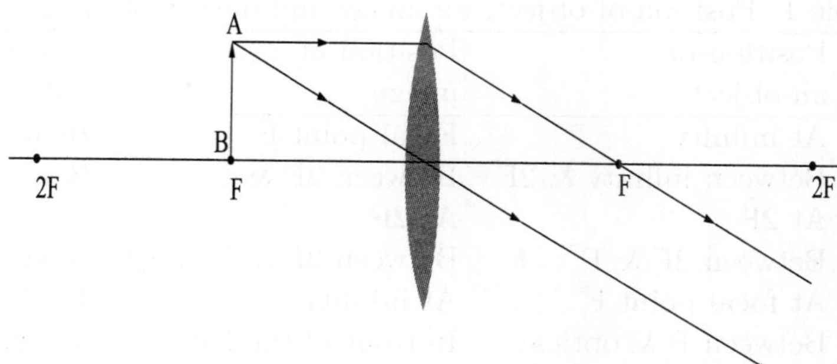

Figure 14: Formation of image when an object is placed at the focal point.

4.6 Object is between the focal point F and optical point

Suppose, an object AB is placed between the focal point F and optical point O (Figure 15). From a point A on the object, a ray parallel to the principal axis of the lens after refraction passes through the focal point F. Other ray from the point A passing through the optical point O goes in the same direction. These two rays from A appear to intersect at A′. This point A′ is virtual image of the point A. Similarly, we can consider other points on the object and get the corresponding images. All these images together form the image A′B′ of the object AB. In this case, the image is formed on the same side of the lens. This image is not real but virtual (imaginary) as it is not formed due to real intersection of refracted rays.

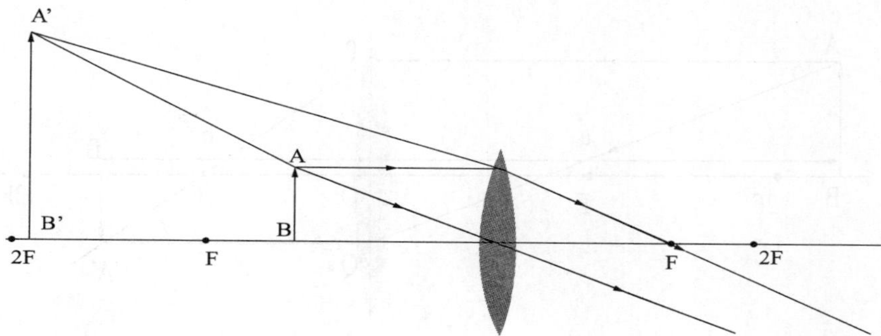

Figure 15: Formation of image when an object is placed between the focal point F and the optical point O.

Note: These six positions of an object, corresponding positions of image of the object and nature of the image in case of a convex lens are summarized in the following Table 1.

Table 1. Position of object, its image and nature of image.

	Position of an object	Position of image	Nature of image
1.	At infinity	Focal point F	Real
2.	Between infinity & 2F	Between 2F & F	Real
3.	At 2F	At 2F	Real
4.	Between 2F & F	Between 2F & infinity	Real
5.	At focal point F	At infinity	Real
6.	Between F & optical point O	In front of the lens	Virtual

4.7 Relation between u, v and f

Here, u and v denote the distances of an object and its image from the optical point O of the lens, respectively. We know that f is the distance of focal point from the optical point which is positive in the case of a convex lens. The relation between u, v and f can be derived in any of the said six cases discussed above of the convex lens. For convenience, let us consider formation of image as shown in Figure 16 where a ray moving parallel to the principal axis of the lens after refraction passes through the focal point F. Another ray going through the focal point F after refraction goes parallel to the principal axis of the lens. These two rays from a point A on the object from the image at the point A$'$. Similarly, we can consider other points on the object and get the corresponding images. All these images together form the image A$'$B$'$ of the object AB.

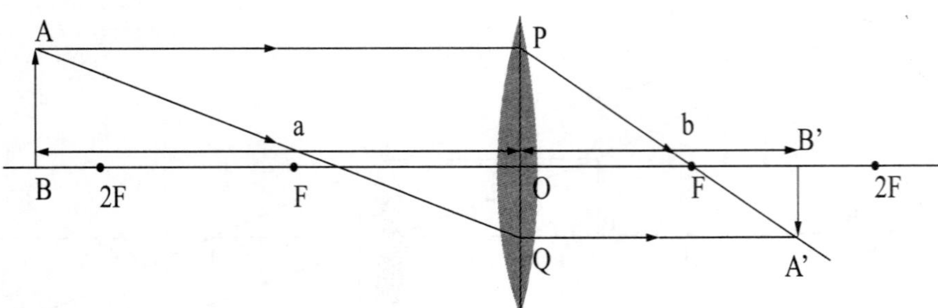

Figure 16: Formation of image A$'$B$'$ of an object AB using a convex lens.

Since AB and OQ are perpendicular to the principal axis, the triangles ABF and OQF are similar. In Δ ABF and Δ OQF, we have

$$\frac{AB}{OQ} = \frac{BF}{FO} \tag{2.1}$$

Since A′B′ and OP are perpendicular to the principal axis, the triangles FOP and FA′B′ are similar. In Δ FOP and Δ FA′B′, we have

$$\frac{OP}{A'B'} = \frac{OF}{FB'} \tag{2.2}$$

Since AB = OP and OQ = A′B′, from equations (2.1) and (2.2), we have

$$\frac{BF}{FO} = \frac{OF}{FB'} \qquad \text{or} \qquad \frac{BO - OF}{FO} = \frac{OF}{OB' - FO} \tag{2.3}$$

Considering the sign convention, we have BO = $-u$, OB′ = v. Now, OF is the focal length which is positive in case of convex lens and we have OF = f. Using these values in equation (2.3), we have

$$\frac{-u - f}{f} = \frac{f}{v - f}$$

On cross multiplication and rearrangement of this equation, we get

$$uf - vf = uv$$

On dividing this equation by uvf, we get

$$\frac{1}{v} - \frac{1}{u} = \frac{1}{f}$$

This is know as the equation of lens. This is the relation between u, v and f where the values are to be used along with the proper sign as per convention.

4.8 Magnification

The ratio of size of image to that of object is known as the magnification, denoted by m. Thus, we have

$$m = \frac{A'B'}{AB}$$

Notice that in Figure 16, the size of image is in the downward direction whereas that of the object in the upward direction. From equation (2.2), we have

$$\frac{A'B'}{OP} = \frac{FB'}{OF}$$

Since AB = OP, we have

$$\frac{A'B'}{AB} = \frac{FB'}{OF} = \frac{OB' - OF}{OF}$$

Considering sign convention, we have $OB' = +b$ and $OF = +f$. Using these values, we have

$$\frac{A'B'}{AB} = \frac{b - f}{f}$$

From the equation of lens, we have

$$\frac{1}{v} - \frac{1}{u} = \frac{1}{f} \qquad \text{or} \qquad \frac{1}{b} + \frac{1}{a} = \frac{1}{f}$$

Here, sign convention is accounted for. Thus, we have $f = ab/(a + b)$. Using this expression for f, we get

$$\frac{A'B'}{AB} = \frac{b - ab/(a + b)}{ab/(a + b)} = \frac{b}{a}$$

Thus, the magnification is

$$m = \frac{b}{a}$$

Thus, magnification of a convex lens is the ratio of the distance of the image to that of the object for the lens.

Exercise 4: An object of length 6 cm is placed at a distance of 40 cm in front of a convex lens of focal length 15 cm. Calculate the position and length of the image.

Solution: We have $u = -40$ cm, $f = +15$ cm. From the equation of lens, we have

$$\frac{1}{v} - \frac{1}{u} = \frac{1}{f}$$

Using the values, we have

$$\frac{1}{v} + \frac{1}{40} = \frac{1}{15}$$

On solving, we get $v = 24$ cm. Thus, a real image is formed at a distance of 24 cm on the right side of the lens. Now, let h' be the height of the image and the size of the object $h = 6$ cm. We have

$$m = \frac{h'}{h} = \frac{24}{40}$$

Using the value of h, we have

$$\frac{h'}{6} = \frac{24}{40} \qquad \text{or} \qquad h' = 3.6 \text{ cm}$$

Thus, an inverted image of 3.6 cm length is formed at a distance of 24 cm from the lens.

Exercise 5: A convex lens of focal length f produces a real image of an object. If the magnification of the image is m and the distance between the the object and image is d, show that

$$f = \frac{md}{(1+m)^2}$$

Solution: Suppose for a convex lens of focal length f, a is the distance of an object and b the distance of its real image as shown in the following figure.

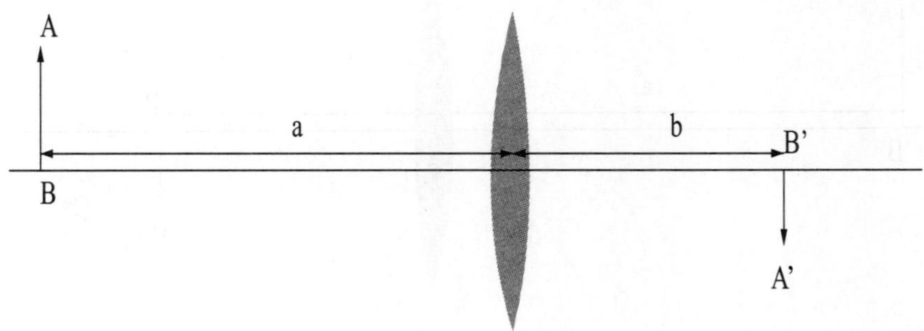

We have magnification m and distance d as

$$m = \frac{b}{a} \qquad \text{and} \qquad d = a + b$$

From the equation of lens, we have

$$\frac{1}{v} - \frac{1}{u} = \frac{1}{f} \qquad \text{or} \qquad \frac{1}{b} + \frac{1}{a} = \frac{1}{f} \qquad (2.4)$$

Multiplying equation (2.4) by b, we have

$$1 + \frac{b}{a} = \frac{b}{f} \qquad \text{or} \qquad 1 + m = \frac{b}{f} \qquad (2.5)$$

Multiplying equation (2.4) by a, we have

$$\frac{a}{b} + 1 = \frac{a}{f} \qquad \text{or} \qquad \frac{1}{m} + 1 = \frac{a}{f} \qquad (2.6)$$

Adding equations (6.12) and (6.13), we get

$$1 + m + \frac{1}{m} + 1 = \frac{b+a}{f} \qquad \text{or} \qquad \frac{m^2 + 2m + 1}{m} = \frac{d}{f}$$

Thus, we have

$$f = \frac{md}{(1+m)^2}$$

Exercise 6: Show that minimum distance between an object and its real image in a convex lens is four times the focal length of the lens.

Solution: Suppose for a convex lens of focal length f, a is the distance of an object and b the distance of its real image as shown in the following figure.

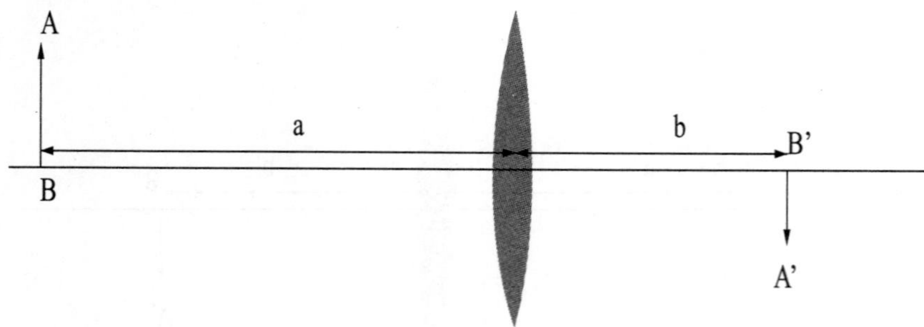

Suppose x is the distance between the object and its image, then we have

$$x = a + b \qquad \text{or} \qquad a = x - b \qquad (2.7)$$

The equation of lens is

$$\frac{1}{v} - \frac{1}{u} = \frac{1}{f} \qquad \text{or} \qquad \frac{1}{b} + \frac{1}{a} = \frac{1}{f} \qquad (2.8)$$

Using the value of a from equations (2.7) in (2.8), we have

$$\frac{1}{b} + \frac{1}{x-b} = \frac{1}{f} \qquad \text{or} \qquad \frac{x-b+b}{b(x-b)} = \frac{1}{f}$$

On simplification, we get

$$b^2 - bx + xf = 0$$

This is a quadratic equation in b and its roots are

$$b = \frac{x \pm \sqrt{x^2 - 4xf}}{2}$$

For b to be real, we should have

$$x^2 - 4xf \geq 0 \qquad \text{or} \qquad x \geq 4f$$

It proves that for the formation of a real image of an object with a convex lens, the minimum distance between the object and its image should be four times the focal length of the lens.

5. Formation of image by a concave lens

In case of a concave lens, the focal point is on the same side as the object, we have therefore, only one case as shown in Figure 17. Here, the image is always formed between the focal point F and the optical point O. The ray from a point A on the object moving parallel to the principal axis of the lens after refraction appears to come from the focal point F of the lens. Second ray from the point A going in the direction of the optical point O will move along the same direction. As these two rays from the point A after refraction appear to intersect at the point A$'$. This A$'$ is the virtual image of A. Similarly, we can consider other points on the object and get the corresponding images. All these images together form the image A$'$B$'$ of the object AB. The image is virtual and on the same side, as the object, of the lens.

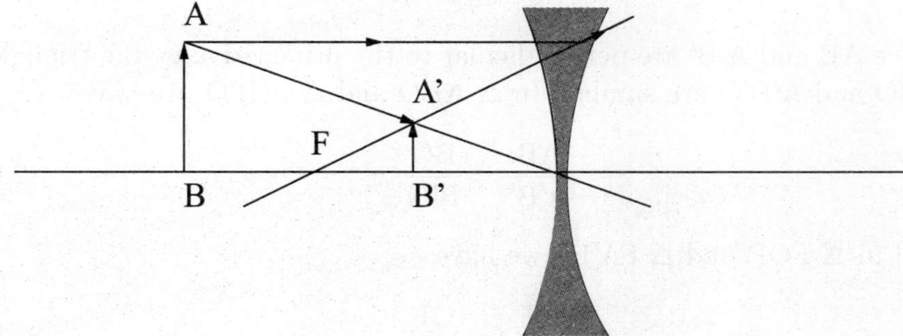

Figure 17: Formation of image A$'$B$'$ of an object AB using a concave lens.

5.1 Relation between u, v and f

Here, u and v denote the distances of an object and its image from the optical point O of the lens, respectively. We know that f is the distance of focal point from the optical point which is negative in the case of a concave lens. For deriving a relation between u, v and f, let us consider Figure 18. A ray moving parallel to the principal axis of the lens after refraction appears to come from the focal point F. Another ray passing through the optical point O goes in the same direction. These two rays from A appear to intersect at A$'$. This point A$'$ is virtual image of the point A. Similarly, we can consider other points on the object and get the corresponding images. All these images together form the image A$'$B$'$ of the object AB.

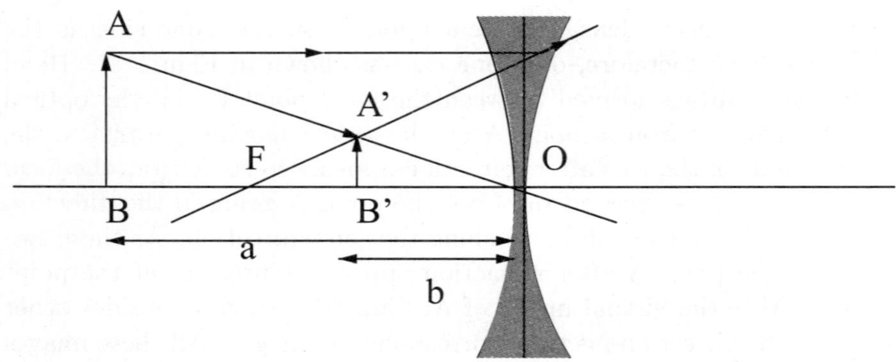

Figure 18: Formation of image A$'$B$'$ of an object AB using a concave lens.

Since AB and A$'$B$'$ are perpendicular to the principal axis, the triangles ABO and A$'$B$'$O are similar. In $\triangle$ ABO and $\triangle$ A$'$B$'$O, we have

$$\frac{AB}{A'B'} = \frac{BO}{B'O} \tag{2.9}$$

and in $\triangle$ FOP and $\triangle$ FA$'$B$'$, we have

$$\frac{OP}{A'B'} = \frac{OF}{FB'} \tag{2.10}$$

Since AB = OP, from equations (2.9) and (2.10), we have

$$\frac{BO}{B'O} = \frac{OF}{FB'} \qquad \text{or} \qquad \frac{BO}{B'O} = \frac{OF}{OF - B'O} \tag{2.11}$$

Considering the sign convention, we have BO $= -u$, OB$' = -v$. Now, OF is the focal length which is negative in case of concave lens and we have OF $= -f$. Using these values in equation (2.11), we have

$$\frac{-u}{-v} = \frac{-f}{-f + v}$$

On cross multiplication and multiplication of this equation, we get

$$uf - vf = uv$$

On dividing this equation by uvf, we get

$$\frac{1}{v} - \frac{1}{u} = \frac{1}{f}$$

This is know as the equation of lens. This is the relation between u, v and f where the values are to be used along with the proper sign as per convention.

5.2 Magnification

The ratio of size of image to that of object is known as the magnification, denoted by m. Thus, we have

$$m = \frac{A'B'}{AB}$$

Notice that in Figure **??**, the size of image as well as that of object are in the upward direction. From equation (2.10), we have

$$\frac{A'B'}{OP} = \frac{FB'}{OF}$$

Since AB $=$ OP, we have

$$\frac{A'B'}{AB} = \frac{FB'}{OF} = \frac{OF - OB'}{OF}$$

Considering sign convention, we have OB$' = -b$ and OF $= -f$. Using these values, we have

$$\frac{A'B'}{AB} = \frac{-f + b}{-f}$$

From the equation of lens, we have

$$\frac{1}{v} - \frac{1}{u} = \frac{1}{f} \qquad \text{or} \qquad -\frac{1}{b} + \frac{1}{a} = -\frac{1}{f}$$

Here, the sign convention is accounted for. Thus, we have $f = ab/(a-b)$. Using this expression for f, we get

$$\frac{A'B'}{AB} = \frac{-ab/(a-b) + b}{-ab/(a-b)} = \frac{b}{a}$$

Thus, the magnification is

$$m = \frac{b}{a}$$

Thus, magnification of a convex lens is the ratio of the distance of the image to that of the object for the lens.

Note: Since the size of image is smaller than that of the object, the magnification of a convex lens is smaller than one. Therefore, a concave is not used magnification purposes.

Exercise 7: An object is placed at a distance of 20 cm in front of a concave lens of focal length 15 cm. Calculate the position of the image.

Solution: We have $u = -20$ cm, $f = -15$ cm. From the equation of lens, we have

$$\frac{1}{v} - \frac{1}{u} = \frac{1}{f}$$

Using the values, we have

$$\frac{1}{v} + \frac{1}{20} = -\frac{1}{15}$$

On solving, we get $v = -8.57$ cm. Thus, an image is formed at a distance of 8.57 cm on the left side of the lens.

6. Power of a lens

Power of a lens is a measure of its ability to produce convergence (convex lens) or divergence (concave lens) of a parallel beam of light. When the focal length f is expressed in meter, the power P is

$$P = \frac{1}{f}$$

As the focal length of a convex lens is positive and that of concave lens negative, the power of a convex lens is positive and that of a concave lens negative. The unit of power is diopter (D).

Exercise 8: Focal length of a convex lens is 80 cm. Calculate power of the lens.

Solution: Focal length of a convex lens is $f = + 80$ cm $= 0.8$ m. The power P of the lens is

$$P = \frac{1}{f} = \frac{1}{0.8} = 1.25 \text{ D}$$

The power of convex lens is $+ 1.25$ D.

Exercise 9: Power of a lens is 1.5 D. Calculate focal length of the lens and write the nature of the lens.

Solution: As the power of the lens is positive, it is a convex (converging) lens. The focal length of a convex lens is

$$f = \frac{1}{D} = \frac{1}{1.5} = 0.6667 \text{ m} = 66.67 \text{ cm}$$

Exercise 10: Focal length of a concave lens is 66 cm. Calculate power of the lens.

Solution: Focal length of a concave lens is $f = -66$ cm $= -0.66$ m. The power P of the lens is

$$P = \frac{1}{f} = \frac{1}{-0.66} = -1.515 \text{ D}$$

The power of concave lens is -1.515 D.

Exercise 11: Power of a lens is $- 0.8$ D. Calculate focal length of the lens and write the nature of the lens.

Solution: As the power of the lens is negative, it is a concave (diverging) lens. The focal length of a concave lens is

$$f = \frac{1}{D} = \frac{1}{-0.8} = -1.25 \text{ m} = -125 \text{ cm}$$

7. Aberrations

One of the basic problems of a lens is the imperfect quality of image formed by it. The deviations in size, shape, position of a colour (wavelength) in the actual image produced by a lens in comparison to the object are known as the aberrations produced by the lens. The aberrations may be classified into two categories: (i) chromatic aberrations and (ii) monochromatic aberrations. The chromatic aberrations are distortions of the image due to the dispersion of light in a lens used in a optical system, when white light is used. Thus, the chromatic aberration may be defined as a defect of coloured image formed by a lens while using white light. When a monochromatic light is used then the chromatic aberrations are automatically absent. Besides the chromatic aberrations, there are some defects which are present even when a monochromatic light is used. Such defects are known as the monochromatic aberrations. These aberrations are due to (i) the large aperture of the lenses, (ii) the large angle subtended by the rays with the principal axis, and (iii) the large size of the object. Examples of monochromatic aberrations are (i) spherical aberration, (ii) astigmatism, (iii) coma, (iv) curvature, and (v) distortion. In the present discussion we shall discuss about the chromatic aberration and spherical aberration.

7.1   Chromatic aberration

The rays of light of different wavelengths (colours) have different velocities in a refracting medium. Thus, a refracting medium has different refractive index for each colour (wavelength). In the visible region of electromagnetic spectrum, the refractive index is the least for red colour and the maximum for the violet colour.[1] It shows that the velocity of red colour is the largest and that of the violet colour the smallest.[2] As the focal length of a lens depends on the refractive index of the medium through the relation

$$\frac{1}{f} = (\mu - 1)\left(\frac{1}{R_1} - \frac{1}{R_2}\right) \tag{2.12}$$

the focal length for the red colour is the largest and that of the violet colour the smallest.[3] Thus, when white light is used a single lens will produce not only one image of the object but one image for each colour (wavelength) at different positions and of different sizes. The variation of the image distance from the lens with refractive index measures the axial or longitudinal chromatic aberration.

---

[1]The refractive index varies with wavelength (colour) as

$$\mu_v > \mu_i > \mu_b > \mu_g > \mu_y > \mu_o > \mu_r$$

[2]The velocity of radiation varies with wavelength (colour) as

$$v_v < v_i < v_b < v_g < v_y < v_o < v_r$$

[3]The focal length varies with wavelength (colour) as

$$f_v < f_i < f_b < f_g < f_y < f_o < f_r$$

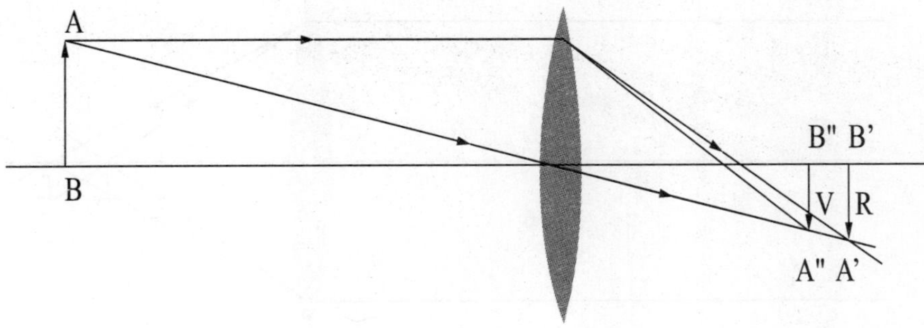

Figure 19: Chromatic aberration.

In the Figure 19 AB is an object illuminated by a white light. Its images corresponding to Red and Violet colours are $A'B'$ and $A''B''$, respectively. The images corresponding to other colours are formed in between these two limiting images. The size of an image depends on the wavelength (colour). Hence, the magnification depends on the wavelength (colour). The distance $x = B''B'$ is the measure of longitudinal chromatic aberration and the distance $y = A'B' - A''B''$ measures the lateral chromatic aberration.

Exercise 12: For a refracting medium, decide if the following statement is correct or not.

(i) The velocity of red colour is larger than that of the violet colour.

(ii) The refractive index of red colour is larger than that of the violet colour.

(iii) Focal length of a lens for red colour is larger than that for the violet colour.

Answer: (i) Yes, (ii) No, (iii) Yes

7.1.1 Chromatic aberration of a lens for an object placed at infinity

The chromatic aberration of a lens for an object placed at infinity can be calculated in the following manner.

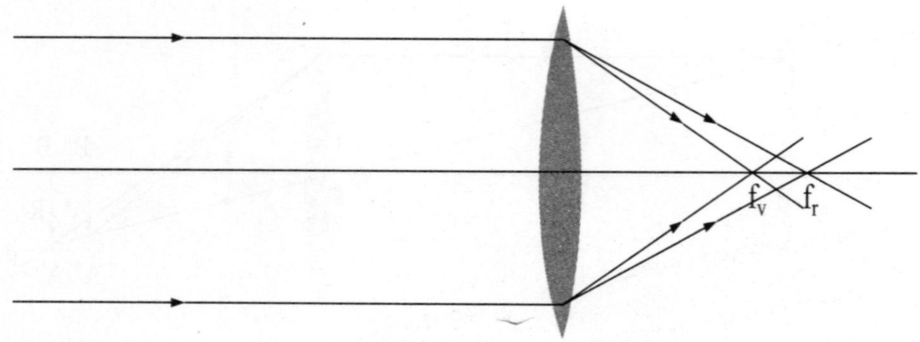

Figure 20: Longitudinal aberration due to convex lens.

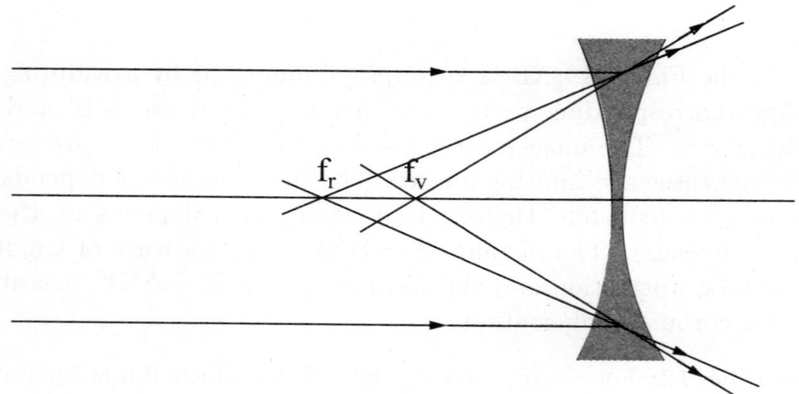

Figure 21: Longitudinal aberration due to concave lens.

Consider the images formed due to longitudinal aberration, as shown in Figures 20 (convex lens) and 21 (concave lens). Let f_r, μ_r and f_v, μ_v be the focal length and refractive index for red and violet colours, respectively. Now, for red colour we have

$$\frac{1}{f_r} = (\mu_r - 1)\left(\frac{1}{R_1} - \frac{1}{R_2}\right) \tag{2.13}$$

and for violet colour we have

$$\frac{1}{f_v} = (\mu_v - 1)\left(\frac{1}{R_1} - \frac{1}{R_2}\right) \tag{2.14}$$

Here, R_1 and R_2 are radii of curvature of two faces of the lens. For the average focal length f and refractive index μ of the lens, we have

$$\frac{1}{f} = (\mu - 1)\left(\frac{1}{R_1} - \frac{1}{R_2}\right) \tag{2.15}$$

From equations (2.13) and (2.15), we have

$$\frac{1}{f_r} = \frac{(\mu_r - 1)}{f(\mu - 1)} \tag{2.16}$$

and from equations (2.14) and (2.15), we have

$$\frac{1}{f_v} = \frac{(\mu_v - 1)}{f(\mu - 1)} \tag{2.17}$$

Subtracting equation (2.16) from (2.17), we get

$$\frac{1}{f_v} - \frac{1}{f_r} = \frac{(\mu_v - \mu_r)}{f(\mu - 1)} = \frac{\omega}{f} \tag{2.18}$$

where $\omega = (\mu_v - \mu_r)/(\mu - 1)$ is known as dispersive power of the material of the lens. Equation (2.18) can be rearranged as

$$\frac{f_r - f_v}{f_r f_v} = \frac{\omega}{f} \qquad \text{or} \qquad f_r - f_v = \frac{\omega}{f} f^2 = \omega f \tag{2.19}$$

where $f = \sqrt{f_r - f_v}$ is the average focal length of the lens. Thus, the longitudinal chromatic aberration $(f_r - f_v)$ is equal to the product of the dispersive power and the mean focal length of the lens, which may be taken as for the yellow light. It is obvious that a single lens with white light cannot produce an image which is free from the chromatic aberration.

7.1.2 Condition for achromatism of two lenses placed in contact

Chromatic aberration may be eliminated by keeping two lenses, made up of different materials, in contact in such a way that all the rays of different colours are brought to focus by the combination at a common point. In practice, the chromatic aberration cannot be removed completely. Usually achromatism is achieved for two prominent colours. A achromatic combination is made by placing in contact two lenses of different materials and of suitable focal lengths so that the focal length of the combination is the same for both the extreme colours, red and violet. The refraction formula for a combination of lens in contact is

$$\frac{1}{f} = (\mu - 1)\left(\frac{1}{R_1} - \frac{1}{R_2}\right) \tag{2.20}$$

where μ is the refractive index of the material of the lens, and R_1 and R_2 are radii of curvatures of two surfaces of the lens. On differentiating equation (2.20), we get

$$-\frac{\mathrm{d}f}{f^2} = \mathrm{d}\mu \left(\frac{1}{R_1} - \frac{1}{R_2}\right) \qquad (2.21)$$

Diving equation (2.21) by (2.20), we get

$$-\frac{\mathrm{d}f}{f^2} = \frac{\mathrm{d}\mu}{f(\mu - 1)} \qquad (2.22)$$

If μ_v and μ_r are the refractive indices for violet and red colours, respectively, then $\mathrm{d}\mu$ is the change in the refractive index such that $\mathrm{d}\mu = \mu_v - \mu_r$. Therefore, the dispersive power of the lens is

$$\omega = \frac{\mathrm{d}\mu}{\mu - 1} \qquad (2.23)$$

Using equation (2.23) in (2.22), we have

$$-\frac{\mathrm{d}f}{f^2} = \frac{\omega}{f} \qquad (2.24)$$

When two lenses of focal lengths f_1 and f_2 are placed in contact, the focal length f of the combination is

$$\frac{1}{f} = \frac{1}{f_1} + \frac{1}{f_2}$$

On differentiating this equation, we get

$$-\frac{\mathrm{d}f}{f^2} = -\frac{\mathrm{d}f_1}{f_1^2} - \frac{\mathrm{d}f_2}{f_2^2}$$

In order that the rays of different colours are brought to focus at a common point, the change in focal length of the combination should be zero, *i.e.*, $\mathrm{d}f = 0$. Thus, we have

$$-\frac{\mathrm{d}f_1}{f_1^2} - \frac{\mathrm{d}f_2}{f_2^2} = 0$$

Using equation (2.24) here, we get

$$\frac{\omega_1}{f_1} + \frac{\omega_2}{f_2} = 0 \qquad (2.25)$$

where ω_1 and ω_2 are dispersive powers for the materials of the lenses, respectively, and are always positive. Equation (2.25) show that f_1 and

f_2 must have opposite signs, *i.e.*, the combination should consist of one concave lens and the other convex lens, generally made up of different materials. Such combination is often known as the achromatic doublet.

Exercise 13: Suppose f_1 and f_2 are focal lengths and ω_1 and ω_2 dispersive powers, respectively, of two lenses placed in contact in an achromat objective of a telescope. If F is the focal length of the objective, show that the focal lengths f_1 and f_2 are

$$f_1 = \frac{(\omega_2 - \omega_1)F}{\omega_2} \qquad \text{and} \qquad f_2 = -\frac{(\omega_2 - \omega_1)F}{\omega_1}$$

Solution: For the achromat objective, we have

$$\frac{\omega_1}{f_1} + \frac{\omega_2}{f_2} = 0 \qquad \text{or} \qquad \frac{1}{f_2} = -\frac{\omega_1}{\omega_2 f_1}$$

We have

$$\frac{1}{F} = \frac{1}{f_1} + \frac{1}{f_2}$$

Therefore, we have

$$\frac{1}{F} = \frac{1}{f_1} - \frac{\omega_1}{\omega_2 f_1} = \frac{\omega_2 - \omega_1}{\omega_2 f_1}$$

Hence,

$$f_1 = \frac{(\omega_2 - \omega_1)F}{\omega_2}$$

and

$$f_2 = -\frac{\omega_2 f_1}{\omega_1} = -\frac{(\omega_2 - \omega_1)F}{\omega_1}$$

Exercise 14: The objective of a telescope is an achromat of focal length 70 cm. When the dispersive powers of two lenses are 0.025 and 0.035, calculate the focal lengths of the two lenses.

Solution: Suppose f_1 and f_2 are focal lengths and ω_1 and ω_2 dispersive powers, respectively, of two lenses, then for an achromat, we have

$$\frac{\omega_1}{f_1} + \frac{\omega_2}{f_2} = 0 \qquad \text{or} \qquad \frac{1}{f_2} = -\frac{\omega_1}{\omega_2 f_1} = -\frac{0.025}{0.035 f_1} = -\frac{5}{7 f_1}$$

The focal length of achromat is

$$\frac{1}{F} = \frac{1}{f_1} + \frac{1}{f_2} = \frac{1}{f_1} - \frac{5}{7 f_1} = \frac{2}{7 f_1}$$

Thus, we have

$$f_1 = \frac{2F}{7} = \frac{2 \times 70}{7} = 20 \text{ cm}$$

and

$$f_2 = -\frac{7}{5}\, f_1 = -\frac{7}{5}\, 20 = -28 \text{ cm}$$

Hence, the focal lengths of the lenses in the objective are $+20$ cm and -28 cm.

7.2 Monochromatic aberrations

The chromatic aberration does not exist when we use monochromatic light. But, another kind of aberrations appear in the picture formed by a lens. The monochromatic light after refraction forms images of an object at different positions and of different sizes. Such aberrations are known as the monochromatic aberrations, which may be of the following types:

(i) Spherical aberration

(ii) Coma

(iii) Astigmatism

(iv) Distortion

7.2.1 Spherical aberration

When a point object O is placed on the principal axis of a large lens, the rays which are away from the axis (marginal rays) are brought to focus at I_m and those near the axis (paraxial rays) are brought to focus at I_p, as shown in Figure 22.

It is clear from Figure 22 that the image formed by the paraxial rays lies at a longer distance than that formed by the marginal rays. The image formed is, therefore, not sharp at any point on the axis. If a screen is placed perpendicular to the axis at I_m, the outer portions of the object are in focus, whereas at I_p, the inner portions of the object are in focus. The spherical aberration for a convex lens is positive. When the screen is placed in between I_m and I_p, the image appears to be a circular patch. This patch is known as the circle of least confusion and corresponds to the position of the best image.

The distance $I_p I_m$ measures the axial or longitudinal spherical aberration. The radius of the circle of least confusion, lying between I_p and I_m, is a measure of the lateral spherical aberration.

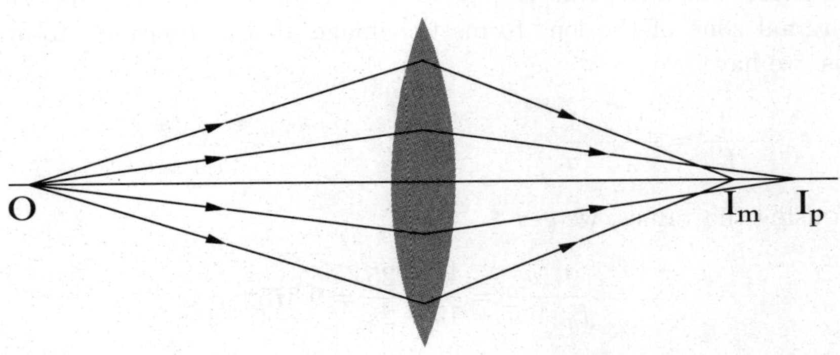

Figure 22: Spherical aberration due to convex lens.

The spherical aberration can be understood in the following manner. A lens can be considered as consists of a number of circular zones, whose axis coincides with the optical axis of the lens. The focal lengths of circular zones vary with the radius of the zone. That is, different zones have different focal lengths. The focal length of marginal zone is smaller than that of paraxial zone. Therefore, the marginal rays are focused faster than the paraxial rays.

7.2.2 Removal of spherical aberration

For a single lens, spherical aberration can be minimized under the following conditions.

(i) By using stops:

The spherical aberration depends on the aperture of lens. It can therefore be reduced by reducing the aperture of the lens with the help of a co-axial aperture stop and therefore using only the central portion of the lens.

(ii) By using a plano-convex lens:

By using a plano-convex lens with its convex surface towards the incident or emergent beam, whichever is more parallel to the lens.

Exercise 15: An object is placed on the optical axis of a convex lens at a distance 25 cm. If the focal length of the marginal and paraxial zones of the lens are 15 cm and 15.3 cm, respectively. Calculate the longitudinal spherical aberration of the lens.

Solution: We have u 25 cm, $f_m = 15$ cm, $f_p = 15.3$ cm. Suppose the marginal zone of the lens forms the image at a distance v_m from the lens, we have

$$\frac{1}{f_m} = \frac{1}{v_m} - \frac{1}{u} \qquad \text{or} \qquad v_m = \frac{f_m u}{f_m + u}$$

On using the values, we get

$$v_m = \frac{f_m u}{f_m + u} = \frac{15 \times 25}{15 + 25} = 9.375 \text{ cm}$$

Similarly, the distance of image from the lens, formed by the paraxial zone is

$$v_p = \frac{f_p u}{f_p + u} = \frac{15.3 \times 25}{15.3 + 25} = 9.491 \text{ cm}$$

Thus, the longitudinal spherical aberration is

$$I_p I_m = v_p - v_m = 9.491 - 9.375 = 0.116 \text{ cm}$$

7.2.3 Coma

As mentioned earlier, the spherical aberration arises due to a point object placed on the axis of the lens. When the object is placed off the axis, the image of such object suffers from an aberration called the coma. This aberration also arises due to different circular zones of the lens. Various zones produce different lateral magnifications in the images. The coma is illustrated in Figure 23, where different zones of the lens, denoted by $(1, 1)$, $(2, 2)$, $(3, 3)$, $(4, 4)$ ... etc. Light rays incident on the lens and after refraction through various zones meet at the screen at different points. For example, the rays passing through $(1, 1)$ zone meet at the point 1. It is found that the rays which incident on the outer zones after refraction through the lens focus near the axis of the lens. Due to circular shape of each zone, the image is circular in shape. Further, the size of the images due to various zones varies and is schematically shown in Figure 23. The result of various images is comet-shaped and therefore the aberration is named as the comatic aberration or coma.

7.2.4 Removal of coma

(i) By using stops:

The coma aberration can be minimized with the help of proper stops placed in front of the lens towards the object side.

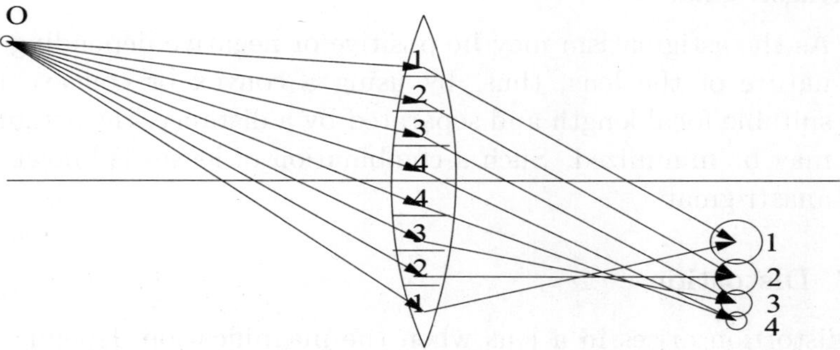

Figure 23: Comatic aberration for a point object.

7.2.5 Astigmatism

In the astigmatism aberration, a lens has different focal lengths across different diameters and therefore produces different images instead of a single image, as shown in Figure 24. The image I_1 formed due to the diameter LM is known as the meridian image and the image I_2 formed due to some other diameter, say RS, is known as the sagittal image. Such defect of image is known as the astigmatism.

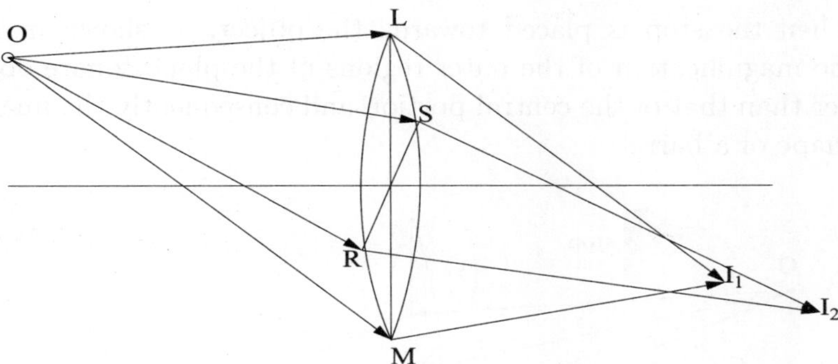

Figure 24: Astigmatism aberration for a point object.

7.2.6 Removal of astigmatism

(i) By using stops:

The astigmatism aberration can be minimized with the help of stops in case of single lens. Stops are placed in such a manner that only less oblique rays pass through the lens to form image.

(ii) Anastrigmat:

As the astigmatism may be positive or negative depending on the nature of the lens, thus, by using a convex or concave lens of suitable focal length and separated by a distance, the astigmatism may be minimized. Such a combination of lenses is known as the anastrigmat.

7.2.7 Distortion

The distortion arises in a lens when the magnification depends on the distance from the principal axis. When the linear magnification produced is different from different parts of the lens then the images of equal parts of an object will be unequal in size. Though the monochromatic aberrations (spherical aberration, coma, astigmatism and curvature) can be reduced by using the stops, the image of a plane square-like object (placed perpendicular to the axis) is not of the shape of a square. This defect is known as the distortion and can be of two types:

1. Barrel-shaped distortion

2. Pin-cushion distortion

Barrel-shaped distortion

When the stop is placed towards the object, as shown in Figure 25, the magnification of the outer regions of the plane square object is smaller than that of the central portion and consequently the image has the shape of a barrel.

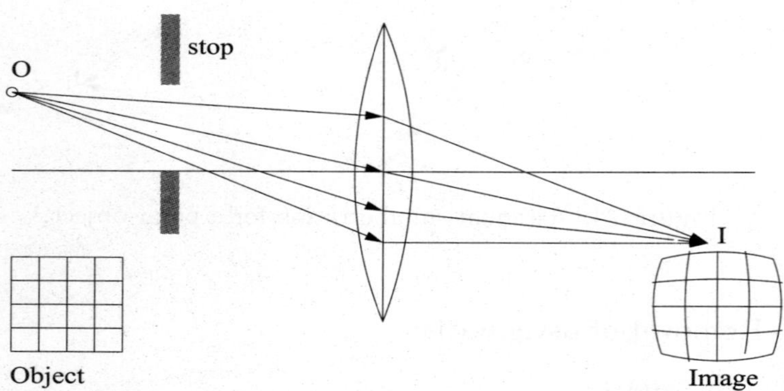

Figure 25: Barrel-shaped distortion.

Pin-cushion distortion

When the stop is placed towards the image, as shown in Figure 26, the magnification of the outer regions of the plane square object is larger than that of the central portion and consequently the image has the shape of a pin-cushion.

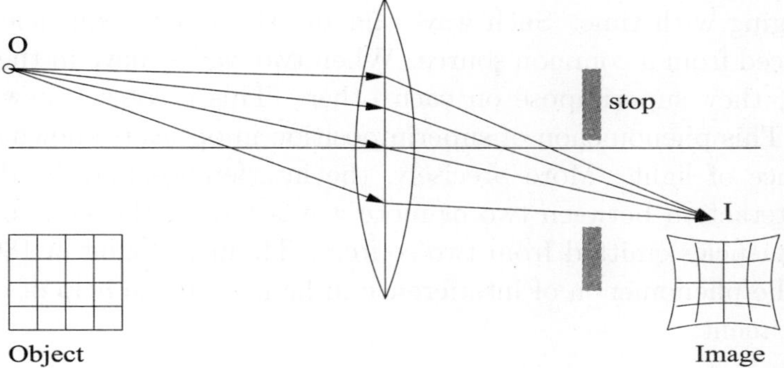

Figure 26: Pin-cushion distortion.

7.2.8 Removal of distortion

For removal of distortion, two meniscus lenses are placed in a manner that their concave surfaces face each other, as shown in Figure 27. Now, place the aperture stop between the two lenses.

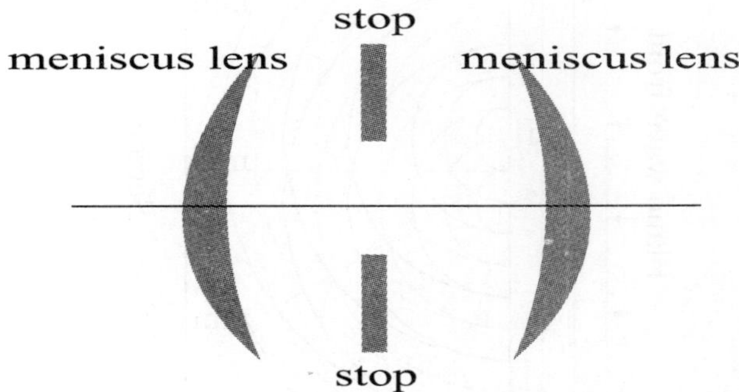

Figure 27: Removal of distortion.

8. Interference

For understanding the phenomenon of interference in optics, we have to account for the wave nature of light. That is the light behaves like a wave. The interference cannot be explained in terms of the linear propagation of light, as discussed in the formation of image of an object with the help of a lens. In the optics, the interference means the superposition of two or more waves. On the superposition of waves, we get a new wave pattern, which is different from those of the waves participating in the superposition. Here, we shall discuss about the interference of waves having the same frequencies, so that the interference pattern is not changing with time. Such waves having the same frequencies may be produced from a common source. When two waves move in the same direction, they superimpose on each other. This creates a new wave pattern. This phenomenon of superimposition in optics is known as the interference of light. More precisely, the interference may be defined as the interaction between two or more waves having the same or very close frequencies emitted from two sources. Thomas Young in 1802 explained the phenomenon of interference in light with the help of double slit experiment.

8.1 Young's double slit experiment

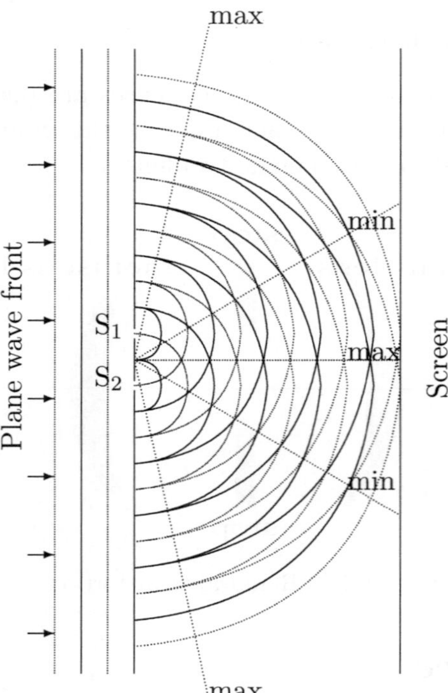

Figure 28: Double slit experiment showing the wave nature of light.

In 1801, Thomas Young experimentally proved that light behaves like a wave. He did so by demonstrating that light undergoes interference as do water waves, sound waves and the waves of all other types. The phenomenon of interference may be understood with the help of Figure 28 where two point sources S_1 and S_2 are derived from a common source. Let the sources S_1 and S_2 are close to each other. Since each source emits waves in all directions, the spherical waves from S_1 and S_2 expand into the space. The crests of the waves are represented by continuous arcs and the troughs by the dotted arcs.

It can be seen that constructive interference takes place at the points where a crest due to one wave meets a crest due to the other wave or where a trough due to one wave meets a trough due to the other wave. In such case, the resulting amplitude will be the sum of amplitudes of two individual waves. Consequently, the intensity of light will be maximum at these points. On the other side, where a crest due to one wave meets a trough due to other wave or vice-versa, the resultant amplitude will be the difference of the amplitudes of individual waves. At these points the intensity of light will be minimum. Thus, due to the interaction of two waves from S_1 and S_2, alternate bright and dark fringes are observed on the screen placed on the right side. These fringes are obtained due to the phenomenon of interference of light.

8.1.1 Path difference and phase difference

Let us consider the phenomenon of interference due to two waves from two sources S_1 and S_2 having the same frequency and the same wavelength. Suppose the sources S_1 and S_2 are separated by a distance d, as shown in Figure 29(a). The interference pattern is formed on the screen. The screen is at a distance D from the slits. Suppose two waves from the sources S_1 and S_2 reach a point P on the screen. When $D >> d$, the waves from sources S_1 and S_2, and reaching at P may be considered to be parallel to each other. If the distances of the sources S_1 and S_2 from P are r_1 and r_2, respectively. The path-difference between the two waves is $\delta = r_2 - r_1$. The phase difference ϕ between the two waves is related to the path-difference δ as

$$\phi = \frac{2\pi}{\lambda}\, \delta$$

where λ is wavelength of the waves. The path-difference between the two waves is [Figure 29(b)]

$$\delta = d \sin \theta \tag{2.26}$$

where d denotes the separation between two sources S_1 and S_2. In Figure 29(a), we have

$$\tan\theta = \frac{y}{D}$$

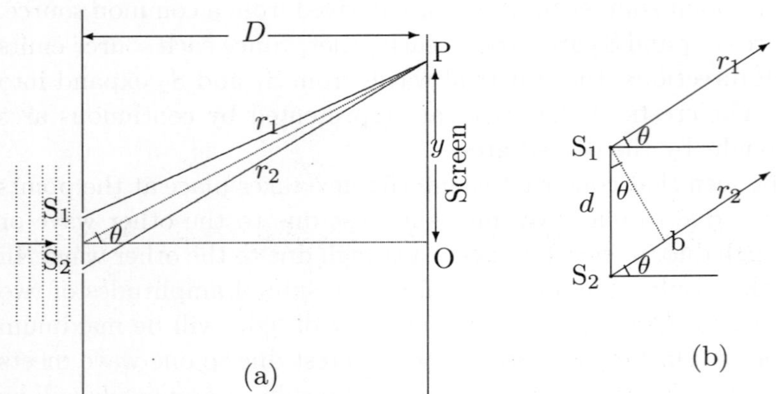

Figure 29: Formation of interference fringes.

As $y << D$, for small angle θ, we have

$$\sin\theta \approx \theta \approx \tan\theta = \frac{y}{D}$$

Therefore, the path-difference expressed in Equation (2.26) is

$$\delta = \frac{yd}{D}$$

8.1.2 Constructive interference

For a bright fringe (constructive interference), the path-difference must be integral multiple of the wavelength. That is,

$$\delta = m\lambda$$

where $m = 0,\ 1,\ 2,\ \ldots$ is a positive integer. For the m-th bright fringe, we have

$$\frac{y_m d}{D} = m\lambda \qquad \text{or} \qquad y_m = \frac{m\lambda D}{d}$$

where y_m denotes the distance of the m-th bright fringe from the mean position O. Hence, the $m = 0$ gives the bright fringe at the central axis. $m = 1$ gives the first order bright fringe; $m = 2$ gives the second order bright fringe; and so on. Thus, the distance of the $(m + 1)$-th bright fringe from the mean position is

$$y_{m+1} = \frac{(m + 1)\lambda D}{d}$$

Thus, the separation between two successive bright fringes is

$$\beta = y_{m+1} - y_m = \frac{(m+1)\lambda D}{d} - \frac{m\lambda D}{d} = \frac{\lambda D}{d}$$

It is also known as the fringe-width.

8.1.3 Destructive interference

For a dark fringe (destructive interference), the path-difference must be half-integer multiple of the wavelength. That is,

$$\delta = \left(m + \frac{1}{2}\right)\lambda$$

where $m = 0, 1, 2, \ldots$ is a positive integer. For the $m+1$-th dark fringe, we have

$$\frac{y_{m+1}d}{D} = \left(m + \frac{1}{2}\right)\lambda \qquad \text{or} \qquad y_{m+1} = \left(m + \frac{1}{2}\right)\frac{\lambda D}{d}$$

where y_{m+1} denotes the distance of the $(m+1)$-th dark fringe from the mean position. Hence, the $m = 0$ gives the first order bright fringe; $m = 1$ gives the second order bright fringe; and so on. Thus, the distance of the $(m+2)$-th dark fringe from the mean position is

$$y_{m+2} = \left(m + 1 + \frac{1}{2}\right)\frac{\lambda D}{d}$$

Thus, the separation between two successive dark fringes is

$$\beta = y_{m+2} - y_{m+1} = \left(m + 1 + \frac{1}{2}\right)\frac{\lambda D}{d} - \left(m + \frac{1}{2}\right)\frac{\lambda D}{d} = \frac{\lambda D}{d}$$

It is also known as the fringe-width. It shows that the separation Δy between two successive bright fringes as well as between two successive dark fringes increases with the increase of the separation D between the screen and the slits. Further, the value of Δy increases with the decreases of the separation d between the two slits, S_1 and S_2.

Exercise 16: A light of wavelength 5500 Å from a narrow slit falls on a double slit. When the separation of 10 fringes on a screen placed 210 cm away from the double slit arrangement is 2.2 cm, calculate the separation between the two slits.

Solution: Given, $D = 210$ cm, wavelength $\lambda = 5500$ Å $= 5.5 \times 10^{-5}$ m and fringe width $\beta = 2.2/10 = 0.22$ cm. The separation d between the two slits is

$$d = \frac{\lambda D}{\beta} = \frac{5.5 \times 10^{-5}(210)}{0.22} = 0.0525 \text{ cm}$$

Exercise 17: Two coherent sources are 0.22 mm apart and the fringes are observed on a screen placed 180 cm away from the sources. It is found that with a monochromatic source of light, the fifth bright fringe is situated at a distance of 30.2 mm from the central fringe.

Solution: Given, $D = 210$ cm, $d = 0.22$ mm $= 0.022$ cm, distance $y = 30.2$ mm $= 3.02$ cm and $n = 10$. The wavelength of light is

$$\lambda = \frac{yd}{nD} = \frac{3.02(0.022)}{5(210)} = 6.328 \times 10^{-5} \text{ cm} = 6.328 \text{ Å}$$

Exercise 18: A source emits light of two wavelengths 4800 Å and 5235 Å. The source is used in the double slit experiment. The interference patterns are observed on a screen placed at a distance 180 cm and the separation between two slits is 0.028 mm. Calculate the separation between the fourth order bright fringes due to these two wavelengths.

Solution: Given, wavelengths $\lambda_1 = 4800$ Å $= 4.8 \times 10^{-5}$ cm and $\lambda_2 = 5235$ Å $= 5.235 \times 10^{-5}$ cm, $n = 4$, $D = 180$ cm, $d = 0.028$ mm $= 0.0028$ cm. Now, we have

$$y_1 = \frac{n\lambda_1 D}{d} \qquad \text{and} \qquad y_2 = \frac{n\lambda_2 D}{d}$$

Thus, the separation between the bright fringes is

$$y = y_2 - y_1 = \frac{n\lambda_2 D}{d} - \frac{n\lambda_1 D}{d} = \frac{nD(\lambda_2 - \lambda_1)}{d}$$

$$= \frac{4(180)(5.235 \times 10^{-5} - 4.8 \times 10^{-5})}{0.0028} = 1.119 \text{ cm}$$

9. Mathematical treatment of interference

Let us consider superposition of two waves of the same angular frequency ω and having a constant phase difference ϕ. Both the waves are traveling along the same direction. Suppose the amplitudes of the waves are A_1 and A_2, respectively. The displacements of the waves may be expressed, respectively, as

$$y_1 = A_1 \, \sin \omega t \tag{2.27}$$

and

$$y_2 = A_2 \, \sin(\omega t + \phi) \tag{2.28}$$

According to the principle of superposition, the resultant displacement y is the algebraic sum of the two displacements. Thus, we have

$$y = y_1 + y_2 = A_1 \sin \omega t + A_2 \sin(\omega t + \phi)$$

$$= A_1 \sin \omega t + A_2[\sin \omega t \cos \phi + \cos \omega t \sin \phi]$$

$$= \sin \omega t (A_1 + A_2 \cos \phi) + \cos \omega t (A_2 \sin \phi)$$

Considering that

$$(A_1 + A_2 \cos \phi) = A' \cos \theta \qquad (2.29)$$

and

$$A_2 \sin \phi = A' \sin \theta \qquad (2.30)$$

we get

$$y = \sin \omega t \, A' \cos \theta + \cos \omega t \, A' \sin \theta = A' \sin(\omega t + \theta)$$

On squaring equations (2.29) and (2.30), and adding them, we get

$$A'^2 = (A_1 + A_2 \cos \phi)^2 + (A_2 \sin \phi)^2 = A_1^2 + A_2^2 + 2A_1 A_2 \cos \phi \quad (2.31)$$

On dividing equations (2.30) by (2.29), we get

$$\tan \theta = \frac{A_2 \sin \phi}{A_1 + A_2 \cos \phi} \qquad (2.32)$$

When $A_1 = A_2 = A$, we have

$$\tan \theta = \frac{\sin \phi}{1 + \cos \phi} = \frac{2 \sin \phi/2 \, \cos \phi/2}{1 + 2\cos^2 \phi/2 - 1} = \tan \phi/2 \qquad (2.33)$$

(i) When $\phi = 2m\pi$ with $m = 0, 1, 2, \ldots$, we have

$$\tan \theta = \frac{\sin 2m\pi}{1 + \cos 2m\pi} = 0 \qquad \text{or} \qquad \theta = 0$$

(ii) When $\phi = (2m + 1)\pi$ with $m = 0, 1, 2, \ldots$, we have

$$\tan \theta = \frac{\sin(2m + 1)\pi}{1 + \cos(2m + 1)\pi} = \frac{0}{0}$$

It is indeterminate. Using the method of dealing with such situation (differentiating numerator and denominator of equation 2.33 separately), we have

$$\tan \theta = \frac{\cos \phi}{- \sin \phi}$$

Putting $\phi = (2m + 1)\pi$, we get

$$\tan \theta = \frac{\cos(2m + 1)\pi}{- \sin(2m + 1)\pi} = \infty \qquad \text{or} \qquad \theta = \pi/2$$

(iii) When $\phi = (4m + 1)\pi/2$ with $m = 0, 1, 2, \ldots$, we have

$$\tan\theta = \frac{\sin[(4m + 1)\pi/2]}{1 + \cos[(4m + 1)\pi/2]} = 1 \qquad \text{or} \qquad \theta = \pi/4$$

(iv) When $\phi = (4m - 1)\pi/2$ with $m = 1, 2, \ldots$, we have

$$\tan\theta = \frac{\sin[(4m - 1)\pi/2]}{1 + \cos[(4m - 1)\pi/2]} = -1 \qquad \text{or} \qquad \theta = 3\pi/4$$

9.1 Intensity of resultant wave

We know that intensity of a wave is proportional to square of its amplitude. Thus, the intensity of the interference pattern is

$$I = A'^2 = A_1^2 + A_2^2 + 2A_1 A_2 \, \cos\phi$$

9.2 Condition for constructive interference

It is obvious from equation (2.31) that the intensity I will be the maximum at the points where the value of $\cos\phi$ is $+1$, *i.e.*, the phase angle ϕ is $2m\pi$ with $m = 0, 1, 2, \ldots$. The maximum intensity is

$$I_{max} = (A_1 + A_2)^2$$

When $A_1 = A_2 = A$, we have $I_{max} = 4A^2$.

9.3 Condition for destructive interference

It is obvious from equation (2.31) that the intensity I will be the minimum at the points where the value of $\cos\phi$ is -1, *i.e.*, the phase angle ϕ is $(2m + 1)\pi$ with $m = 0, 1, 2, \ldots$. The minimum intensity is

$$I_{min} = (A_1 - A_2)^2$$

When $A_1 = A_2 = A$, we have $I_{min} = 0$.

Exercise 19: Two monochromatic sources having same frequency and intensity ratio 16:1 is producing interference fringes. Calculate the ratio of intensities of maximum and minimum in the fringe system.

Solution: Suppose the amplitudes of two waves participating in the interference are A_1 and A_2. Their intensities respectively are $I_1 = A_1^2$ and $I_2 = A_2^2$. We are given $I_1 : I_2 = 16 : 1$. Hence, $A_1^2 : A_2^2 = 16 : 1$. Thus, we have

$$\frac{A_1^2}{A_2^2} = \frac{16}{1} \qquad \text{or} \qquad A_1 = 4A_2$$

In the interference pattern, the ratio of maximum intensity I_{max} and minimum intensity I_{min} is

$$\frac{I_{max}}{I_{min}} = \frac{(A_1 + A_2)^2}{(A_1 - A_2)^2}$$

Using $A_1 = 4A_2$, we have

$$\frac{I_{max}}{I_{min}} = \frac{(4A_2 + A_2)^2}{(4A_2 - A_2)^2} = \frac{25}{9}$$

Therefore, $I_{max} : I_{min} = 25 : 9$.

Exercise 20: Two monochromatic sources having same frequency and amplitudes 5 and 4 cm produce interference fringes. Calculate the ratio of intensities of maximum and minimum in the fringe system.

Solution: Given, the amplitudes $A_1 = 5$ cm and $A_2 = 4$ cm. The ratio of maximum and minimum intensities in the interference pattern is

$$\frac{I_{max}}{I_{min}} = \frac{(A_1 + A_2)^2}{(A_1 - A_2)^2} = \frac{(5 + 4)^2}{(5 - 4)^2} = \frac{81}{1}$$

9.4 Average energy for interference

When two waves having the same frequency ω interfere, constructive interference and destructive interference are obtained regularly. For the waves (2.27) and (2.28), the maximum and minimum intensities, respectively, are expressed by equations (2.32) and (2.33). The average intensity is

$$I_{av} = \frac{I_{max} + I_{min}}{2}$$

Using the expressions for I_{max} and I_{min}, we have

$$I_{av} = \frac{(A_1 + A_2)^2 + (A_1 - A_2)^2}{2} = A_1^2 + A_2^2$$

When $A_1 = A_2 = A$, we have $I_{av} = 2A^2$.

9.5 Variation of intensity

When $A_1 = A_2 = A$, the intensity is

$$I = A^2 + A^2 + 2A^2 \cos \phi = 2A^2(1 + \cos \phi) = 4A^2 \cos^2 \phi/2$$

The variation of intensity I as a function of ϕ in the interference is shown in Figure 30 for the case $A_1 = A_2 = A$. The intensity varies between 0 and $4A^2$.

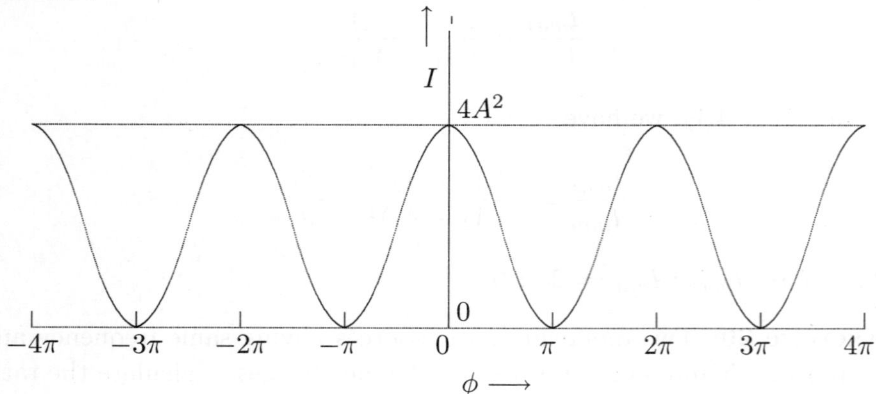

Figure 30: Variation of intensity I as a function of ϕ in the interference.

9.6 Conditions for sustained interference

We naturally would like to have a sustained interference which does not vary with time. In order to obtain such interference, the following conditions need to be satisfied.

(i) The two sources participating in the interference should be monochromatic.

(ii) The two sources should emits the waves having the same frequency (wavelength)

(iii) The waves from the sources should propagate along the same direction with equal speed.

(iv) The phase difference between the waves should not depend on time.

(v) The two sources should be kept very close to each other. For large separation between the sources, clear maxima and minima in the pattern will not appear.

(vi) A reasonable distance between the sources and the screen should be kept, as the maxima and minima appear quite close when the distance is small. However, a large distance reduces the intensity.

(vii) In order to obtain distinct maxima and minima with good contrast, the amplitudes of the two interfering waves should be equal.

10. Interference in a uniform thin film

Consider a uniform transparent thin film having thickness t and refractive index $\mu > 1$. The upper surface of the film is PQ and the lower is RS. A wave of light AB incident on the upper surface PQ of the film at an angle i. At the point B, the wave is partly reflected along BC and partly refracted along BF. The angle of refraction at B is r. Inside the film, the wave moves along BF and incident at the lower surface RS at the point F. At F, the angle of incidence is r. The wave at F is partly reflected along FD and partly refracted along FK. The angle of refraction at F is i. The phenomena of reflection and refraction at the surfaces PQ and RS will go on. In this situation, the interference occurs between the waves BC and DE (above the film), and between the waves FK and GL (below the film). The waves going in the upward direction from the surface PQ are generally known as the reflected waves, and the waves going in the downward direction from the surface RS are generally known as the transmitted waves. We draw perpendicular DM from the point D on BC and perpendicular GJ from the point G on FK.

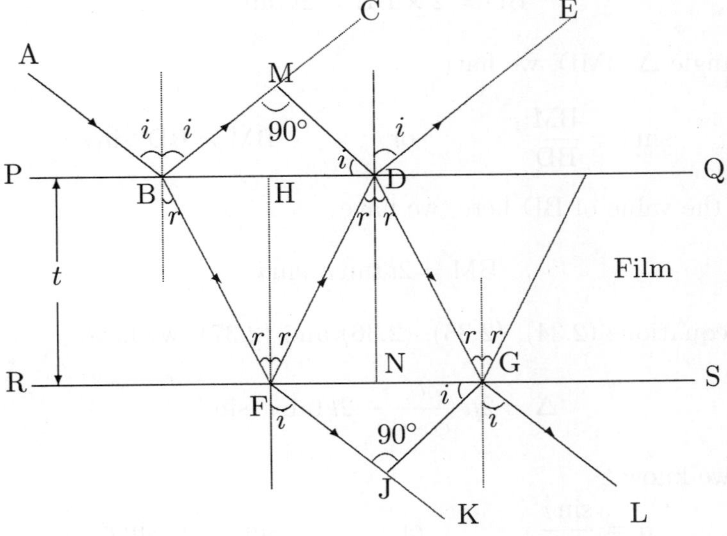

Figure 31: Interference in a uniform thin film.

10.1 Interference between reflected waves

Let us consider two reflected waves BC and DE. Beyond the line MD, both the waves travel equal distance. The path-difference between the two waves is introduced during the journey from B to D through F and from B to M. Thus, the path-difference is

$$\Delta = (\text{BF} + \text{FD})_{\text{in film}} - (\text{BM})_{\text{in air}}$$

$$= \mu(\text{BF} + \text{FD}) - \text{BM}$$

Since BF = FD, we have

$$\Delta = 2\mu\text{BF} - \text{BM} \tag{2.34}$$

In triangle Δ BFH, we have

$$\cos r = \frac{t}{\text{BF}} \qquad \text{or} \qquad \text{BF} = \frac{t}{\cos r} \tag{2.35}$$

and

$$\tan r = \frac{\text{BH}}{t} \qquad \text{or} \qquad \text{BH} = t\tan r$$

Therefore,

$$\text{BD} = 2 \times \text{BH} = 2t\tan r \tag{2.36}$$

In triangle Δ BMD, we have

$$\sin i = \frac{\text{BM}}{\text{BD}} \qquad \text{or} \qquad \text{BM} = \text{BD}\ \sin i$$

Using the value of BD here, we have

$$\text{BM} = 2t\tan r\ \sin i \tag{2.37}$$

From equations (2.34), (2.35), (2.36) and (2.37), we have

$$\Delta = 2\mu\ \frac{t}{\cos r} - 2t\tan r\ \sin i \tag{2.38}$$

Now, we know

$$\mu = \frac{\sin i}{\sin r} \qquad \text{or} \qquad \sin i = \mu\ \sin r \tag{2.39}$$

Using equation (2.39) in (2.38), we have

$$\Delta = \frac{2\mu t}{\cos r} - 2t\ \frac{\sin r}{\cos r}\ \mu\ \sin r$$

$$= \frac{2\mu t}{\cos r}\ (1 - \sin^2 r) = 2\mu t\ \cos r \tag{2.40}$$

Equation (2.40) represents only the apparent path difference and does not represent the effective path difference. When the light is reflected from a surface of an optically denser medium, a phase change which is equivalent to the path-difference of $\lambda/2$ is introduced. Thus, the total path-difference is

$$\Delta = 2\mu t \, \cos r + \lambda/2$$

Obviously, the path-difference depends on the thickness t of the film. It should be noted that the interference pattern will not be perfect as the intensities of the rays BC and DE are not the same and their amplitude are different.

10.1.1 Condition for maxima

To have a maximum at a given point, the two waves should arrive at the point in phase. That is, the path-difference should be integer multiple of the wavelength λ of the wave. Therefore, we have

$$\Delta = n\lambda$$

where n can assume integer values 0, 1, 2, ... Thus, we get

$$2\mu t \, \cos r + \lambda/2 = n\lambda \qquad \text{or} \qquad 2\mu t \, \cos r = (2n - 1)\lambda/2$$

10.1.2 Condition for minima

To have a minimum at a given point, the two waves should arrive at the point completely out of phase. That is, the path-difference should be half-integer multiple of the wavelength λ of the wave. Therefore, we have

$$\Delta = (n + 1/2)\lambda$$

where n can assume integer values 0, 1, 2, ... Thus, we get

$$2\mu t \, \cos r + \lambda/2 = (2n + 1)\lambda/2 \qquad \text{or} \qquad 2\mu t \, \cos r = n\lambda$$

10.2 Interference between transmitted waves

Let us consider two transmitted waves FK and GL. Beyond the line GJ, both the waves travel equal distance. The path-difference between the two waves is introduced during the journey from F to G through D and from F to J. Thus, the path-difference is

$$\Delta = (FD + DG)_{\text{in film}} - (FJ)_{\text{in air}}$$

$$= \mu(\text{FD} + \text{DG}) - \text{FJ}$$

Since FD = DG, we have

$$\Delta = 2\mu\text{FD} - \text{FJ} \tag{2.41}$$

In triangle Δ FDN, we have

$$\cos r = \frac{t}{\text{FD}} \qquad \text{or} \qquad \text{FD} = \frac{t}{\cos r} \tag{2.42}$$

and

$$\tan r = \frac{\text{FN}}{t} \qquad \text{or} \qquad \text{FN} = t \tan r$$

Therefore,

$$\text{FG} = 2 \times \text{FN} = 2t \tan r \tag{2.43}$$

In triangle Δ FJG, we have

$$\sin i = \frac{\text{FJ}}{\text{FG}} \qquad \text{or} \qquad \text{FJ} = \text{FG} \ \sin i$$

Using the value of FG here, we have

$$\text{FJ} = 2t \tan r \ \sin i \tag{2.44}$$

From equations (2.41), (2.42), (2.43) and (2.44), we have

$$\Delta = 2\mu \ \frac{t}{\cos r} - 2t \tan r \ \sin i \tag{2.45}$$

Now, we know

$$\mu = \frac{\sin i}{\sin r} \qquad \text{or} \qquad \sin i = \mu \ \sin r \tag{2.46}$$

Using equation (2.46) in (2.45), we have

$$\Delta = \frac{2\mu t}{\cos r} - 2t \ \frac{\sin r}{\cos r} \ \mu \ \sin r = \frac{2\mu t}{\cos r} \ (1 - \sin^2 r) = 2\mu t \ \cos r \tag{2.47}$$

Equation (2.47) represents the effective path difference. For reflection at the rarer medium, additional path-difference is not introduced.

10.2.1 Condition for maxima

To have a maximum at a given point, the two waves should arrive at the point in phase. That is, the path-difference should be integer multiple of the wavelength λ of wave. Therefore, we have

$$\Delta = n\lambda$$

where n can assume integer values 0, 1, 2, ... Thus, we get

$$2\mu t \ \cos r = n\lambda$$

10.2.2 Condition for minima

To have a minimum at a given point, the two waves should arrive at the point completely out of phase. That is, the path-difference should be half-integer multiple of the wavelength λ of the wave. Therefore, we have

$$\Delta = (n + 1/2)\lambda$$

where n can assume integer values 0, 1, 2, ... Thus, we get

$$2\mu t \,\cos r = (2n + 1)\lambda/2$$

Here also, the interference pattern will not be perfect as the intensities of the waves FK and GL are not the same and their amplitude are different. It should be noted the conditions for the interference between reflected waves are opposite to those for the interference between transmitted waves. It shows that the interference patterns between reflected waves and between transmitted waves are complimentary to each other.

Exercise 21: Light of 5893 Å is reflected at normal incidence from a soap film of refractive index $\mu = 1.4$. Calculate the least thickness of the film that will appear (i) bright and (ii) dark.

Solution: For normal incidence ($i = 0$ and $r = 0$) on a thin film, bright fringe is observed when

$$2\mu t = (2n - 1)\lambda/2 \qquad \text{or} \qquad t = \frac{(2n - 1)\lambda}{4\mu}$$

The least thickness for bright fringe is

$$t = \frac{\lambda}{4\mu} = \frac{5893 \times 10^{-10}}{4 \times 1.4} = 1.05 \times 10^{-7} \text{ m}$$

The minimum thickness of film to see a bright fringe is 1.05×10^{-7} m. For normal incidence on a thin film, dark fringe is observed when

$$2\mu t = n\lambda \qquad \text{or} \qquad t = \frac{n\lambda}{2\mu}$$

The least thickness for dark fringe is

$$t = \frac{\lambda}{2\mu} = \frac{5893 \times 10^{-10}}{2 \times 1.4} = 2.1 \times 10^{-7} \text{ m}$$

The minimum thickness of film to see a dark fringe is 2.1×10^{-7} m.

Exercise 22: Calculate the smallest thickness of a soap film (refractive index $\mu = 1.46$) that will result in a constructive interference in the reflected light when the film is illuminated normally with light whose wavelength is 5893 Å.

Solution: Given $\lambda = 5893 \times 10^{-10}$ m and $\mu = 1.46$. For normal incidence, the thickness of soap film for constructive interference is

$$t = \frac{(2n-1)\lambda}{4\mu}$$

The smallest thickness of the soap film for constructive interference is

$$t = \frac{\lambda}{4\mu} = \frac{5893 \times 10^{-10}}{4 \times 1.46} = 1.0 \times 10^{-7} \text{ m}$$

Exercise 23: A wave of light of wavelength 5893 Å incident on a thin glass plate (refractive index $\mu = 1.5$) such that the angle of refraction in the plate is 60°. Calculate the smallest thickness of the glass plate which will appear dark in the reflected light.

Solution: Given $\lambda = 5893 \times 10^{-10}$ m, $\mu = 1.5$ and the angle of refraction $r = 60°$. The thickness of soap film for destructive interference is

$$2\mu t \, \cos r = n\lambda \qquad \text{or} \qquad t = \frac{n\lambda}{2\mu \, \cos r}$$

The smallest thickness of the glass plate is

$$t = \frac{\lambda}{2\mu \, \cos r} = \frac{5893 \times 10^{-10}}{2 \times 1.5 \times \cos 60°} = 3.93 \times 10^{-7} \text{ m}$$

Exercise 24: The rhinestones in costume jewelry are glass with index of refraction 1.50. To make them more reflective, they are coated with a layer of silicon monoxide of index of refraction 2.00. What is the minimum coating thickness needed to ensure that light of wavelength 560 nm and perpendicular incidence will be reflected from the two surfaces of the coating with fully constructive interference?

Solution: For a thin film (coating) of thickness t and index of refraction μ, the condition for constructive interference is

$$2\mu t = (n + 1/2)\lambda \qquad \text{or} \qquad t = \frac{(n + 1/2)\lambda}{2\mu}$$

For minimum thickness, we have $m = 0$ and

$$t = \frac{\lambda}{4\mu} = \frac{560 \times 10^{-9}}{4 \times 2} = 7 \times 10^{-8} \text{ m}$$

11. Coherence

A wave which appears to be a pure sine-wave for an infinitely long period of time, or in an infinitely extended space, is known as a perfectly coherent wave. For such a wave, there is definite relationship between the phase of the wave at a given time and at a certain time later, or at a given point and at a point which is at a certain distance away. No actual light source, however, emits a perfectly coherent wave. Light waves which are pure sine-waves for a limited period of time or in a limited space are partially coherent waves. There are two different criterion of coherence: (i) the criterion of time, called the temporal coherence and (ii) the criterion of space, called the spatial coherence.

11.1 Temporal coherence

Here, we discuss about the variation with time. While the phase of the oscillating electric field E of radiation would vary linearly with time, a perfectly coherent light wave would have a constant amplitude of vibration at any given time. A time variation of electric field would appear as an ideal sinusoidal wave (Figure 32).

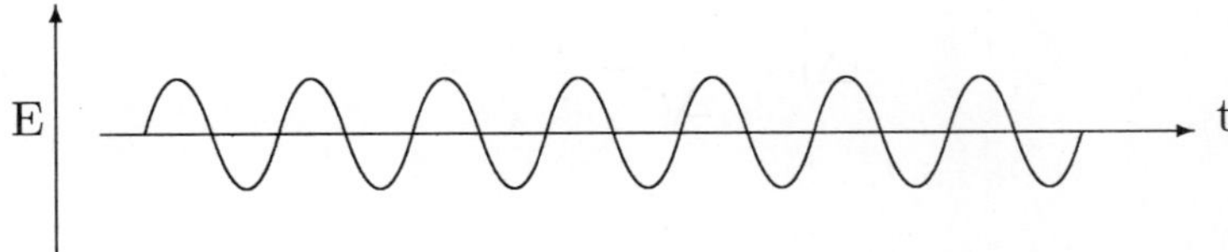

Figure 32: Variation of E with time in an ideal sinusoidal wave.

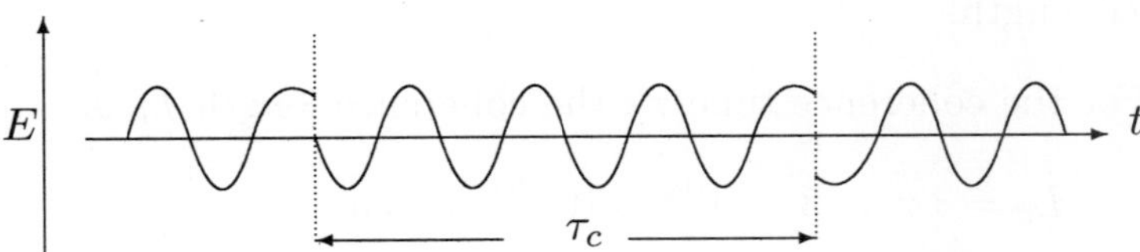

Figure 33: Temporal coherence over the time τ_c.

However, no light emitted by an actual source produces an ideal sinusoidal field at all values of time. This is because when an excited atom returns to the lower state, it emits a pulse of short duration, such as of the order of 10^{-10} s for sodium atom, for example. Thus, the field remains sinusoidal for the time-intervals of the order of 10^{-10} s, after that the phase changes abruptly. Thus, the field due to an actual light source will be as shown in Figure 33.

The average time-interval for which the field remains sinusoidal (*i.e.*, till the definite phase relationship exists) is known as the time of coherence, and is denoted by τ_c. The distance L_c for which the field is sinusoidal is expressed as

$$L_c = c\, \tau_c$$

where c is the speed of light and L_c is known as the coherence length of the light beam. The coherence time is

$$\tau_c = \frac{L_c}{c}$$

According to Fourier analysis, the frequency band width $\Delta\nu$ is related to the average life-time Δt of the upper level of transition, as

$$\Delta\nu = \frac{1}{\Delta t}$$

However, Δt is the time during which a wave train is radiated by the source atom and corresponds to the coherence time τ_c and therefore, we have

$$\Delta\nu = \frac{1}{\Delta t} = \frac{1}{\tau_c} \tag{2.48}$$

and thus,

$$\Delta\nu = \frac{c}{L_c} \tag{2.49}$$

Exercise 25: A sodium lamp has a coherence time $\sim 10^{-10}$ s, calculate the coherence length.

Solution: For the coherence time τ_c, the coherence length L_c is

$$L_c = c\, \tau_c = 3 \times 10^{10} \times 10^{-10} = 3 \text{ cm}$$

11.2 Spatial coherence

Here, we discuss about the variation with space. The spatial coherence is the phase relationship betweeen the radiation fields at different points in space. In order to understand the spatial coherence, let us consider light waves emitted from a point source S (Figure 34). Suppose A and B be two points lying on a line joining them with S. The phase relationship between the points A and B depends on the distance AB and on the temporal coherence of the beam. When AB $< L_c$, where L_c is the coherence length, there will be a definite phase relationship between the points A and B. On the other side, when AB $> L_c$, there will be no coherence between the points A and B.

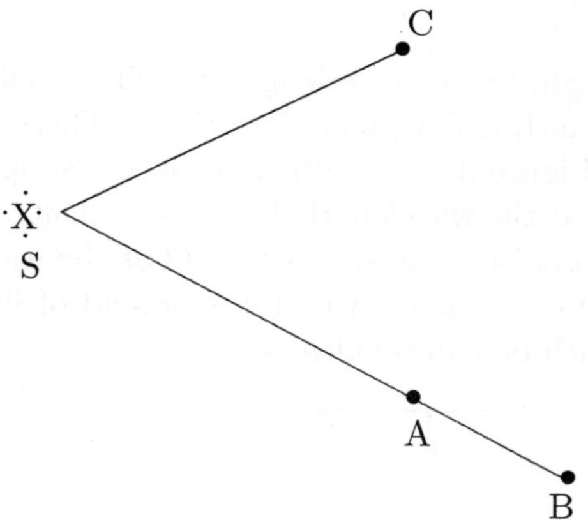

Figure 34: Diagram for explaining the spatial coherence.

Let us now consider two points A and C, which are equidistant from the point source S. Then the waves from S will reach the points A and C in exactly the same phase, that is, the two points will have perfect (spatial) coherence. However, in case of an extended source, the points A and C will not be in coherence. Thus, the spatial coherence is related to the size of the source.

Exercise 26: Suppose a light of 6800 Å has wave trains of 12.5 $\times 10^{-6}$ m long. Calculate the coherence time.

Solution: Given, $\lambda = 6800 \times 10^{-10}$ m, coherence length $L_c = 12.5 \times 10^{-6}$ m. Therefore, the coherence time is

$$\tau_c = \frac{L_c}{c} = \frac{12.5 \times 10^{-6}}{3 \times 10^8} = 4.17 \times 10^{-14} \text{ s}$$

Exercise 27: Coherence length of a light is 3.6 $\times 10^{-4}$ m and wavelength 5500 Å. Calculate the coherence time and number of oscillations corresponding to the coherence length.

Solution: Given, $\lambda = 5500 \times 10^{-10}$ m, coherence length $L_c = 3.6 \times 10^{-4}$ m. The coherence time is

$$\tau_c = \frac{L_c}{c} = \frac{3.6 \times 10^{-4}}{3 \times 10^8} = 1.2 \times 10^{-12} \text{ s}$$

The number of oscillations in the coherence length L_c is

$$n = \frac{L_c}{\lambda} = \frac{3.6 \times 10^{-4}}{5500 \times 10^{-10}} = 654.5$$

11.3 Line-width

Consider monochromatic light having wavelength λ_0. Though it is monochromatic light, but its wavelength is not precisely defined. There is continuous spread, as shown in Figure 35, on both sides of λ_0. Suppose, I_{max} be the maximum intensity at the wavelength λ_0. The intensity decreases on both sides of λ_0. Suppose, the intensity is $I_{max}/2$ at the wavelengths λ_1 and λ_2. Then, $\Delta\lambda = \lambda_2 - \lambda_1$ is known as the spread of light. It is also known as the line-width or band-width.

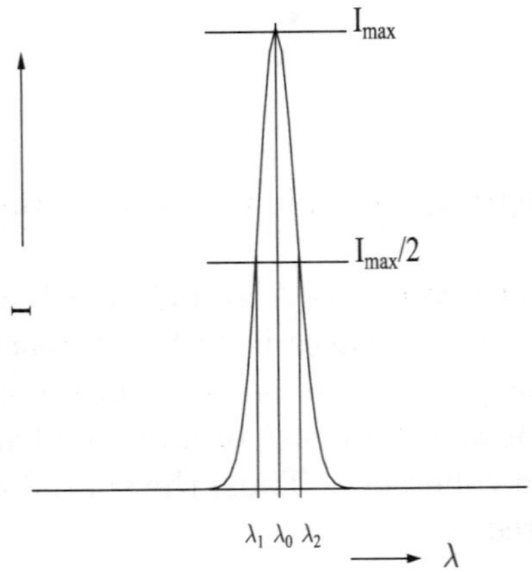

Figure 35: Variation of intensity I with wavelength λ.

11.4 Non-monochromacity

If ν_0 be the central frequency of monochromatic line with line-width $\Delta\nu$, the non-monochromacity ξ is expressed as

$$\xi = \frac{\Delta\nu}{\nu_0}$$

It shows that with the increase of line-width, the non-monochromacity increases.

11.5 Relation between coherence length and band-width

The frequency ν and wavelength λ of a radiation is related as

$$\nu = \frac{c}{\lambda}$$

Here, c is the speed of light. On differentiation of this relation, we have

$$\Delta\nu = -\frac{c}{\lambda^2}\Delta\lambda$$

Using relation (2.49) here, we get

$$\frac{c}{L_c} = -\frac{c}{\lambda^2}\Delta\lambda \qquad \text{or} \qquad L_c = \frac{\lambda^2}{\Delta\lambda}$$

The minus sign has been dropped as it has no significance. This is relation between the coherence length L_c and the band-width $\Delta\lambda$. it shows that the coherence length is inversely proportional to the line-width.

Exercise 28: For a Krypton line having wavelength 6100 Å and coherence length 20 cm, calculate the line-width, coherence time and non-monochromacity of the line.

Solution: Given $\lambda = 6100$ Å $= 6.1 \times 10^{-7}$ m, $L_c = 20$ cm $= 0.2$ m. The frequency ν of radiation is

$$\nu = \frac{c}{\lambda} = \frac{3 \times 10^8}{6.1 \times 10^{-7}} = 4.918 \times 10^{14} \text{ Hz}$$

The coherence time is

$$\tau_c = \frac{L_c}{c} = \frac{0.2}{3 \times 10^8} = 6.67 \times 10^{-10} \text{ s}$$

The line-width is

$$\Delta\nu = \frac{c}{L_c} = \frac{3 \times 10^8}{0.2} = 1.5 \times 10^9 \text{ Hz}$$

The non-monochromacity of line is

$$\xi = \frac{\Delta\nu}{\nu} = \frac{1.5 \times 10^9}{4.918 \times 10^{14}} = 3.05 \times 10^{-6}$$

Exercise 29: For a Krypton line having wavelength 6100 Å the Doppler width is 5.5×10^{-13} m. Calculate the coherence length of the line.

Solution: Given, $\lambda = 6100$ Å $= 6.1 \times 10^{-7}$ m, $\Delta\lambda = 5.5 \times 10^{-13}$ m. The coherence length is

$$L_c = \frac{\lambda^2}{\Delta\lambda} = \frac{(6.1 \times 10^{-7})^2}{5.5 \times 10^{-13}} = 0.6765 \text{ m}$$

Exercise 30: Calculate the coherence length of a laser beam having the band-width 3500 Hz.

Solution: Given, the band-width $\Delta\nu = 3500$ Hz. The coherence time τ_c is

$$\tau_c = \frac{1}{\Delta\nu} = \frac{1}{3500} = 2.857 \times 10^{-4} \text{ s}$$

The coherence length is

$$L_c = c\tau_c = 3 \times 10^8 (2.857 \times 10^{-4}) = 8.57 \times 10^4 \text{ m}$$

Exercise 31: For a red cadmium light of wavelength 6438 Å with coherence length 42 cm, calculate the coherence time and line width.

Solution: Given, coherence length $L_c = 42$ cm $= 0.42$ m and wavelength $\lambda = 6438$ Å $= 6.438 \times 10^{-7}$ m. The coherence time is

$$\tau_c = \frac{L_c}{c} = \frac{0.42}{3 \times 10^8} = 1.4 \times 10^{-9} \text{ s}$$

The line width is

$$\Delta\lambda = \frac{\lambda^2}{L_c} = \frac{(6.438 \times 10^{-7})^2}{0.42} = 9.87 \times 10^{-13} \text{ m}$$

12. Diffraction

For obtaining the Fraunhoffer diffraction, the incident wave front must be plane and the diffracted waves are collected on a screen with the help of an achromatic lens. Hence, the source of light should be either at a large distance from the slit or a collimation lens must be used.

12.1 Fraunhoffer diffraction at a single slit

Let us consider a plane wave front of monochromatic light of wave length λ incident on a rectangular slit AB of width b, as shown in Figure 36. The slit is placed perpendicular to the plane of the paper. We can divide this wave front AB into a large number of points, each producing a large number of secondary waves of equal amplitude a. These secondary waves get diffracted and then interfere to produce a diffraction pattern on the screen. The secondary waves traveling along the direction of the incident beam are focused at the point C whereas those inclined at an angle θ with the direction of the incident beam are focused at a point P.

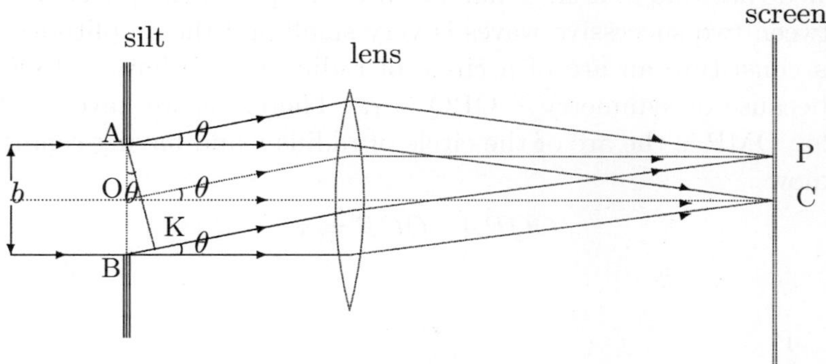

Figure 36: Fraunhoffer diffraction by a single slit.

In order to find out the resultant intensity at the point P, we draw a perpendicular AK on BK. Figure 36 shows that the optical paths of all the waves traveled after the plane AK to the point P are equal. However, the optical paths of the waves produced from the points on the slit AB (from A towards B) and reaching at the point P increases gradually. Thus, the phase difference between them gets larger, as shown in Figure 37(a) for 4 parts, each of amplitude a. This figure has a phase difference of ϕ between two successive waves. The resultant magnitude of these waves at the point P is denoted by R. Here, $\alpha = 4\phi$.

The total path difference between the wave originating for the extreme points A and B of the silt is $BK = AB \sin\theta = b\sin\theta$. The path difference between the waves originating from various points on AB vary from zero to $b\sin\theta$. The value is zero for the wave originated from the point A and $b\sin\theta$ for the wave originated from the point B. The phase difference corresponding to the path difference $b\sin\theta$ is

$$\frac{2\pi}{\lambda}\, b\sin\theta$$

When the slit AB is divided into n equal parts, the phase difference between the waves generated from two successive parts is

$$\frac{1}{n}\,\frac{2\pi}{\lambda}\, b\sin\theta = \phi \ \text{(say)}$$

The resultant amplitude and intensity at the point P due to all these secondary waves can be obtained with the help of vector polygon method. Suppose α be the phase difference between the initial direction and the direction of the resultant, as shown in Figure 37(a). Then, 2α

will be total phase difference between the secondary waves originating from two extreme points A and B, as shown in Figure 37(b). Here, it is assumed that due to a large number n of the parts, the phase difference ϕ between two successive waves is very small and the amplitudes of the waves constitute an arc of a circle of radius r. We have $\angle$ POQ $= \alpha$ and because of symmetry $\angle$ OPQ $= \alpha$. Therefore, we have $\angle$ OQP $= \pi - 2\alpha$. OMP is the arc of the circle of radius r and having center at C. We know

$$\angle OQP + \angle OCP = \pi$$

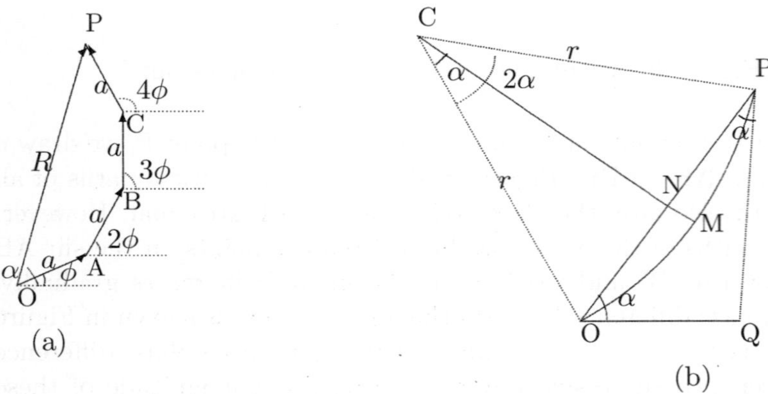

Figure 37: (a) is the vector polygon for $n = 4$. When the value of n is very large, the polygon is like an arc of a circle.

Therefore, $\angle$ OCP $= 2\alpha$. The chord OP gives the resultant amplitude at the point P due to all the secondary waves. In Δ OCN, we have

$$\sin \alpha = \frac{\text{ON}}{\text{OC}} = \frac{\text{ON}}{r} \qquad \text{or} \qquad \text{ON} = r \sin \alpha$$

The chord ONP $= 2$ ON $= 2r \sin \alpha$. Since the chord OP is the resultant amplitude R, we have

$$R = 2r \sin \alpha \qquad\qquad (2.50)$$

The length of the arc OMP $= na$ where n is the number of equal parts of AB and a the amplitude of each secondary wave, originated from a point on AB. We know that

$$\angle \text{PCO} = \frac{\text{arc OMP}}{\text{radius}} \qquad 2\alpha = \frac{na}{r} \qquad 2r = \frac{na}{\alpha} \qquad (2.51)$$

Using the value of $2r$ from equation (2.51) in (2.50), we have

$$R = \frac{na}{\alpha} \sin \alpha \qquad \text{or} \qquad R = A \frac{\sin \alpha}{\alpha}$$

where $A = na$. The resultant intensity at the point P is

$$I = R^2 = I_0 \left(\frac{\sin \alpha}{\alpha}\right)^2$$

Here, $I_0 = A^2$ represents the intensity at $\theta = 0$. It shows that the magnitude of the resultant intensity at a point P depends on α and therefore on the slit width b. As the phase difference of 2α is introduced due to the path difference of $b \sin \theta$, we have

$$2\alpha = \frac{2\pi}{\lambda} b \sin \theta \qquad \text{or} \qquad \alpha = \frac{\pi}{\lambda} b \sin \theta$$

It shows that α depends on the angle of diffraction θ and therefore, $\sin^2 \alpha / \alpha^2$ gives the intensity at different values of θ.

12.1.1 Position of central (principal) maximum

For the central point C on the screen, we have $\theta = 0$ and therefore, $\alpha = 0$. Thus, we have

$$\lim_{\alpha \to 0} \frac{\sin \alpha}{\alpha} = 1$$

and the intensity at C is $I = I_0$.

12.1.2 Position of minima

The intensity is minimum when

$$\frac{\sin \alpha}{\alpha} = 0$$

such that $\sin \alpha = 0$, but $\alpha \neq 0$. Therefore, we have

$$\alpha = \pm m\pi \qquad \text{where } m = 1, 2, 3, \ldots$$

Hence, we have

$$\frac{\pi}{\lambda} b \sin \theta = \pm m\pi \qquad \text{or} \qquad b \sin \theta = \pm m\lambda$$

This is the condition for m-th order minimum.

12.1.3 Secondary maxima

The direction of m-th order maxima is expressed as

$$b \sin \theta = \pm \left(m + \frac{1}{2} \right) \lambda$$

This is equivalent to

$$\alpha = \frac{\pi}{\lambda} \, b \sin \theta = \pm \frac{\pi}{\lambda} \left(m + \frac{1}{2} \right) \lambda = \pm \left(m + \frac{1}{2} \right) \pi$$

Therefore, we have

$$\alpha = \pm \frac{3\pi}{2}, \ \pm \frac{5\pi}{2}, \ \pm \frac{7\pi}{2}, \dots$$

where the secondary maxima are situated. The intensity of the first secondary maximum is

$$I_1 = I_0 \left(\frac{\sin 3\pi/2}{3\pi/2} \right)^2 = \frac{4}{9\pi^2} \, I_0$$

The intensity of the second secondary maximum is

$$I_2 = I_0 \left(\frac{\sin 5\pi/2}{5\pi/2} \right)^2 = \frac{4}{25\pi^2} \, I_0$$

12.1.4 Ratio of intensities of secondary maxima

The ratio of intensities of the successive maxima is

$$I_0 : \frac{4}{9\pi^2} \, I_0 : \frac{4}{25\pi^2} \, I_0 : \frac{4}{49\pi^2} \, I_0 : \dots \qquad \text{or} \qquad 1 : \frac{4}{9\pi^2} : \frac{4}{25\pi^2} : \frac{4}{49\pi^2} : \dots$$

12.1.5 Width of central maximum

Suppose y is the distance of first minimum from the center of the principal maximum. The width of the central maxima is $w = 2y$. For small angle θ, we have

$$\tan \theta = \sin \theta = \frac{y}{D}$$

The condition for first minimum is

$$b \sin \theta = \pm \lambda \qquad \text{or} \qquad \sin \theta = \pm \frac{\lambda}{b}$$

Thus, we have

$$\frac{y}{D} = \pm\frac{\lambda}{b} \qquad \text{or} \qquad y = \pm\frac{\lambda D}{b}$$

Thus, the width of the central maxima is

$$w = 2y = \frac{2\lambda D}{b}$$

It shows that the size of the central maximum increases with the increase of distance of screen from the slit, and with the decrease of the width of slit.

12.1.6 Effect of the slit width

We have

$$\sin\theta = \pm\frac{\lambda}{b}$$

(i) When b is large, $\sin\theta$ is small, and consequently, the angle θ is small. It shows that the maxima and minima would be close to the central maximum.

(ii) When b is small, $\sin\theta$ is large, and consequently, the angle θ is large. It shows that the maxima and minima would be quite distinct and clear.

Exercise 32: A light of wavelength 5500 Å falls normally on a slit of width 2.3 μ m. Calculate the angular displacement of second and third minima.

Solution: Given, $b = 2.3 \times 10^{-6}$ m and $\lambda = 5500$ Å $= 5.5 \times 10^{-7}$ m. For the second minima, the angular displacement θ is expressed as

$$\sin\theta = \frac{2\lambda}{b} = \frac{2 \times 5.5 \times 10^{-7}}{2.3 \times 10^{-6}} = 0.4783$$

or

$$\theta = \sin^{-1}(0.4783) = 28.57°$$

For the third minima, the angular displacement θ is expressed as

$$\sin\theta = \frac{3\lambda}{b} = \frac{3 \times 5.5 \times 10^{-7}}{2.3 \times 10^{-6}} = 0.7174$$

or

$$\theta = \sin^{-1}(0.7174) = 45.84°$$

The angular displacements for second and third order maxima are 28.57° and 45.84°.

Exercise 33: In the Fraunhoffer diffraction, a narrow slit of width 2.4 μ m is illuminated by monochromatic light of wavelength 5600 Å. Calculate the half angular width of the central maximum.

Solution: Given, $b = 2.4 \times 10^{-6}$ m and $\lambda = 5600$ Å $= 5.6 \times 10^{-7}$ m. The first minimum in the diffraction pattern is in the direction of the half angular width of the central maximum. Thus, we have

$$\sin\theta = \frac{m\lambda}{b} = \frac{5.6 \times 10^{-7}}{2.4 \times 10^{-6}} = 0.2333$$

or

$$\theta = \sin^{-1}(0.2333) = 13.49°$$

Thus, the half-angular width of the central maximum is 13.49°.

Exercise 34: In the Fraunhoffer diffraction, a beam of monochromatic light of wavelength 5600 Å incident normally on a slit. The central maximum fans out at 15° on both side of the central position. Calculate the width of the slit.

Solution: Given, $\lambda = 5600$ Å $= 5.6 \times 10^{-7}$ m, $\theta = 15°$ for the first minimum. The width of the slit is

$$b = \frac{\lambda}{\sin\theta} = \frac{5.6 \times 10^{-7}}{\sin 15°} = \frac{5.6 \times 10^{-7}}{0.2588}$$

$$= 2.16 \times 10^{-6} \text{ m} = 2.16 \ \mu\text{m}$$

Exercise 35: In the Fraunhoffer diffraction, a beam of monochromatic light of wavelength 5700 Å incident normally on a slit of width 0.1 mm. Calculate the angular displacement and linear size of central maximum formed on a screen placed 120 cm away from the slit.

Solution: Given, $b = 0.1$ mm $= 0.1 \times 10^{-3}$ m and $\lambda = 5700$ Å $= 5.7 \times 10^{-7}$ m, $x_0 = 120$ cm $= 1.2$ m. For first minimum, we have

$$\sin\theta = \frac{\lambda}{b} = \frac{5.7 \times 10^{-7}}{0.1 \times 10^{-3}} = 5.7 \times 10^{-3}$$

or

$$\theta = \sin^{-1}(5.7 \times 10^{-3}) = 0.3265°$$

Thus, total spread of central maximum is

$$2\theta = 2 \times 0.3265 = 0.653°$$

For linear width z of the half of the central maximum is

$$z = D \, \sin \theta = 1.2 \times 5.7 \times 10^{-3} = 6.84 \times 10^{-3} \text{ m}$$

Therefore, total linear width of the central maximum is

$$2z = 2 \times 6.84 \times 10^{-3} = 1.368 \times 10^{-3} \text{ m}$$

Exercise 36: In the Fraunhoffer diffraction, a slit is illuminated by a light having two wavelengths λ_1 and λ_2. It is found that the first minimum obtained for λ_1 coincides with the second minimum obtained for λ_2. Find out the relation between λ_1 and λ_2.

Solution: For the first minimum due to λ_1, we have

$$b \sin \theta = \lambda_1$$

For the second minimum due to λ_2, we have

$$b \sin \theta = 2\lambda_2$$

Therefore, we have

$$\lambda_1 = 2\lambda_2$$

Exercise 37: Calculate angular width of central bright maximum in the Fraunhoffer diffraction pattern of a slit of width 2.6 μ m illuminated by a monochromatic light of wavelength 5800 Å.

Solution: Given, $b = 2.6 \times 10^{-6}$ m and $\lambda = 5800$ Å $= 5.8 \times 10^{-7}$ m. For first minimum, we have

$$\sin \theta = \frac{\lambda}{b} = \frac{5.8 \times 10^{-7}}{2.6 \times 10^{-6}} = 0.223 \qquad \text{or} \qquad \theta = \sin^{-1}(0.223) = 12.89°$$

Therefore, total angular width of central maximum is

$$2\theta = 2 \times 12.89 = 25.78°$$

12.2 Fraunhoffer diffraction at double slit

Suppose a parallel, collimated beam of monochromatic light of wavelength λ be incident normally on two parallel slits AB and DE as shown in Figure 38. For convenience, we consider the slits of equal width b ($=$ AB $=$ DE) and separated by an opaque distance d ($=$ BD). The distance between two corresponding points of two slits is $(b + d)$. Suppose the

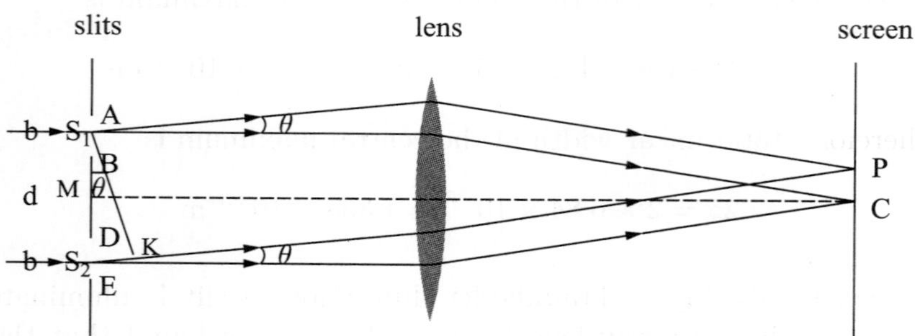

Figure 38: Fraunhoffer diffraction at double slit.

diffracted light is focused with the help of a convex lens on the screen placed in the focal plane of the lens. Obviously, all the secondary waves produced at the slits S_1 and S_2, and propagating parallel to MC (the direction of incident wave) get focused at a point C. Hence, the point C corresponds to the position of the central bright maximum.

Here, we consider that the two slits are equivalent to two coherent sources placed at the middle points S_1 and S_2 of the slits AB and DE, respectively. From the discussion of single slit, we know that the amplitude R' due to each slit of width b is

$$R' = \frac{A \sin \alpha}{\alpha} \tag{2.52}$$

at a point P on the screen, making an angle θ with MC. Here,

$$\alpha = \frac{\pi b \sin \theta}{\lambda}$$

and A is a constant. We may consider that each slit is sending a wave of amplitude $A \sin \alpha / \alpha$ and these two waves will produce interference pattern. The resultant amplitude due to interference having a phase difference of ϕ can be calculated in the following manner. Draw a perpendicular S_1K from the point S_1 on S_2K. Thus, the path difference between two rays reaching at the point P is

$$S_2K = (b + d) \sin \theta$$

The phase difference between them is

$$\phi = \frac{2\pi}{\lambda} (b + d) \sin \theta \tag{2.53}$$

The resultant amplitude R at the point P can be determined with the help of the vector addition method as shown in Figure 39. Since both the slits are of the same size and therefore send the waves of the same amplitude R', we cam write

$$\vec{R} = \vec{R}' + \vec{R}'$$

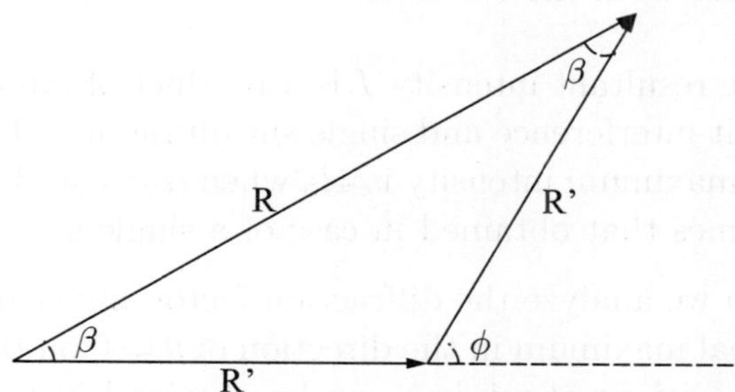

Figure 39: Vector addition method for finding resultant amplitude. R' and R' are amplitudes of the waves from the two slits and ϕ the phase difference between the two waves.

On squaring this equation, we get

$$R^2 = R'^2 + R'^2 + 2\,\vec{R}'\cdot\vec{R}' = 2R'^2 + 2R'^2 \cos\phi$$

$$= 2R'^2(1 + \cos\phi) = 4R'^2 \cos^2(\phi/2) \tag{2.54}$$

From Figure 39, we have

$$\phi = \beta + \beta \qquad \text{or} \qquad \beta = \phi/2 \tag{2.55}$$

From equations (3.1) and (2.55), we have

$$\beta = \frac{\pi}{\lambda}\,(b + d)\sin\theta$$

Using equations (2.52) and (2.55) in (2.54), we get

$$R^2 = 4A^2\,\frac{\sin^2\alpha}{\alpha^2}\,\cos^2\beta$$

Thus, the intensity I is expressed as

$$I = R^2 = 4I_0\,\frac{\sin^2\alpha}{\alpha^2}\,\cos^2\beta \tag{2.56}$$

where $I_0 = A^2$. Equation (2.56) shows that the resultant intensity depends on two factors:

(i) The factor $I_0 \sin^2 \alpha / \alpha^2$, which gives the diffraction pattern due to a single slit.

(ii) The factor $\cos^2 \beta$, which gives the interference pattern due to the waves from the two slits.

Thus, the resultant intensity I is a product of intensities obtained from double slit interference and single slit diffraction. Equation (2.56) shows that the maximum intensity is $4I_0$ when $\alpha = 0$ and $\beta = 0$. This intensity is four times that obtained in case of a single slit.

When we analyze the diffraction factor $\sin^2 \alpha / \alpha^2$, we find that there is principal maximum in the direction of $\theta = 0$ on the screen at the point C. The positions of minima can be obtained in the direction $\sin \alpha = 0$ when $\alpha \neq 0$. Therefore,

$$\alpha = m\pi \qquad \frac{\pi b \sin \theta}{\lambda} = m\pi \qquad b \sin \theta = m\lambda \qquad (2.57)$$

where $m = \pm 1, \pm 2, \pm 3, \ldots$.   The factor $\sin^2 \alpha / \alpha^2$ gives secondary maxima at the points where

$$\alpha = \frac{(2m+1)\pi}{2} \qquad \text{or} \qquad b \sin \theta = \frac{(2m+1)}{2} \lambda$$

where $m = \pm 1, \pm 2, \pm 3, \ldots$. Notice that $|\alpha| \neq \pi/2$ as the first minimum is at $|\alpha| = \pi$.

When we analyze the variation of intensity due to the factor $\cos^2 \beta$. It give the minimum intensity when

$$\cos^2 \beta = 0 \qquad \cos \beta = 0 \qquad \beta = \frac{(2n+1)\pi}{2}$$

where $n = 0, \pm 1, \pm 2, \pm 3, \ldots$. Using expression for β, we get

$$\frac{\pi}{\lambda} (b+d) \sin \theta = \frac{(2n+1)\pi}{2} \qquad \text{or} \qquad (b+d) \sin \theta = \frac{(2n+1)\lambda}{2}$$

For the maximum intensity in the interference factor, we have

$$\cos^2 \beta = 1 \qquad \beta = n\pi$$

where $n = 0, \pm 1, \pm 2, \pm 3, \ldots$. Using expression for β, we get

$$\frac{\pi}{\lambda} (b+d) \sin \theta = n\pi \qquad \text{or} \qquad (b+d) \sin \theta = n\lambda$$

12.2.1 Missing orders of diffraction pattern

In the diffraction pattern due to a double slit, the width of each slit is b and the separation between the slits as d. When the width b is kept constant, the diffraction pattern remains the same. Keeping the value of b as constant and varying the value of d, the position of the interference maxima changes. Depending on the relative values of d and b, certain orders of interference maxima will be missing in the resultant pattern. The directions of interference maxima are expressed as

$$(b+d)\sin\theta = n\lambda \tag{2.58}$$

where n is an integer. The directions of diffraction minima are expressed as

$$b\sin\theta = m\lambda \tag{2.59}$$

where m is an integer. When the values of d and b are such that both the equations (2.58) and (2.59) are satisfied simultaneously for the same value of θ, then the positions of certain interference maxima will correspond to the diffraction minima at the same position on the screen. Those interference maxima will be missing in the diffraction pattern. On dividing equation (2.58) by (2.59), we get

$$\frac{b+d}{b} = \frac{n}{m}$$

(i) Let $b = d$, then we have

$$\frac{d+d}{d} = \frac{n}{m} \qquad \text{or} \qquad n = 2m$$

Therefore, for

$$m = 1, 2, 3, \ldots \qquad \text{we have} \qquad n = 2, 4, 6, \ldots$$

Hence, the orders 2, 4, 6, ... of the interference maxima will be missing in the diffraction pattern.

(ii) Let $d = 2b$, then we have

$$\frac{2b+b}{b} = \frac{n}{m} \qquad \text{or} \qquad n = 3m$$

Therefore, for

$$m = 1, 2, 3, \ldots \qquad \text{we have} \qquad n = 3, 6, 9, \ldots$$

Hence, the orders 3, 6, 9, ... of the interference maxima will be missing in the diffraction pattern.

(iii) Let $d = 0$, then the two silts join and form a single slit of width $2b$. Moreover, we have $n = m$. Now, all the orders of the interference maximum will be missing.

Exercise 38: In Fraunhoffer diffraction due to a narrow slit, a pattern is obtained on a screen placed 2 m away from the slit. The slit width is 0.2 mm and the first minima lie 5 mm away from the central maximum on either side. Find the wavelength of monochromatic light used in the experiment.

Solution: We have $b = 0.2$ mm $= 0.02$ cm, $y = 5$ mm $= 0.5$ cm, $x_0 = 2$ m $= 200$ cm. In case of Fraunhoffer diffraction at a narrow slit, the minima are expressed as

$$b \sin\theta = m\lambda$$

For the first minimum $(m = 1)$, we have

$$\sin\theta = \frac{\lambda}{b}$$

We have

$$\sin\theta = \frac{z}{x_0}$$

Therefore, we have

$$\frac{z}{x_0} = \frac{\lambda}{b} \qquad \text{or} \qquad \lambda = \frac{zb}{x_0}$$

Using the values, we get

$$\lambda = \frac{zb}{x_0} = \frac{0.02 \times 0.5}{200} = 5 \times 01^{-5} \text{ cm} = 5000 \text{ Å}$$

Exercise 39: In Fraunhoffer diffraction due to a narrow slit, a pattern is obtained on a screen placed at a large distance from the slit. The first minima on either side of central maximum is observed in the direction making an angle of $\pm 12°$. For the monochromatic light of 5500 Å used in the experiment, calculate the width of the slit.

Solution: For the first minimum in the Fraunhoffer diffraction due to a narrow slit, we have

$$b \sin\theta = \lambda \qquad \text{or} \qquad b = \frac{\lambda}{\sin\theta}$$

Using the values, we get

$$b = \frac{\lambda}{\sin\theta} = \frac{5500 \times 10^{-8}}{\sin 12°} = \frac{5500 \times 10^{-8}}{0.2079} = 2.65 \times 10^{-4} \text{ cm} = 2.65 \ \mu\text{m}$$

Exercise 40: In the Fraunhoffer diffraction due to a narrow slit, calculate the relative intensities of the first, second and third secondary maxima.

Solution: In the Fraunhoffer diffraction due to a narrow slit, the intensity of secondary maxima is expressed as

$$I = I_0 \left(\frac{\sin^2 \alpha}{\alpha^2}\right)^2 \qquad \text{where} \qquad \alpha = \frac{(2n+1)\pi}{2}$$

For the first, second and third secondary maxima, we have, respectively,

$$\alpha = \frac{3\pi}{2}, \qquad \alpha = \frac{5\pi}{2}, \qquad \alpha = \frac{7\pi}{2}$$

Relative to the intensity I_0 of the central maximum, the intensity of the first secondary maximum is

$$\frac{I_1}{I_0} = \left(\frac{\sin 3\pi/2}{3\pi/2}\right)^2 = 4.50 \times 10^{-2}$$

Relative to I_0, the intensity of the second secondary maximum is

$$\frac{I_2}{I_0} = \left(\frac{\sin 5\pi/2}{5\pi/2}\right)^2 = 1.62 \times 10^{-2}$$

Relative to I_0, the intensity of the third secondary maximum is

$$\frac{I_3}{I_0} = \left(\frac{\sin 7\pi/2}{7\pi/2}\right)^2 = 8.27 \times 10^{-3}$$

Exercise 41: Calculate missing orders in a double slit Fraunhoffer diffraction pattern when the width of each slit is 0.06×10^{-5} m and they are 0.03×10^{-5} m apart.

Solution: The directions of interference maxima are expressed as

$$(b + d)\sin\theta = n\lambda$$

where n is an integer. The directions of diffraction minima are expressed as

$$b\sin\theta = m\lambda$$

where m is an integer. For missing maxima, we have

$$\frac{b+d}{b} = \frac{n}{m}$$

Using the values of b and d, we have

$$\frac{b+d}{b} = \frac{n}{m} \qquad \text{or} \qquad \frac{0.03 \times 10^{-5} + 0.06 \times 10^{-5}}{0.06 \times 10^{-5}} = 1.5 = \frac{n}{m}$$

Thus, we have $n = 1.5$ m.

12.3 Diffraction grating

The Fraunhoffer diffraction at two slits consists of diffraction maxima and minima given by $\sin^2 \alpha/\alpha^2$ and the interference maxima and minima in each diffraction maximum governed by the factor $\cos^2 \beta$. We have also seen that the maximum intensity produced by a double slit is four times larger than that of single slit. Therefore, it is expected that a device having large number of slits will produce large intensity.

Let us consider a device having N slits. For convenience, we consider that each of the N slits is of equal width b and they are separated by equal opaque part of size d. Such a device is known as the transmission grating. It can be constructed by ruling a large number of fine, equidistant lines on a transparent plane glass plate with the help of fine diamond point. The ruled lines are opaque to light whereas the space between any two lines is transparent to the light and acts as a slit. There are about 5000 lines per cm or more in such a grating.

Suppose a parallel, collimated beam of monochromatic light of wavelength λ incident normally on diffraction grating having N slits, each of width b and separated by a opaque distance d. The factor $(b + d)$ is known as the grating element. Obviously, in two successive slits, the middle points are separated by $(b+d)$. Let the diffracted light is focused by a convex lens on the screen placed in the focal plane of the lens. All the secondary waves from the slits traveling in the direction parallel to the direction of incident wave are brought to focus at the point C (Figure 40), which corresponds to the position of the central bright maximum. The waves making an angle θ with the direction of incidence are focused at a point P.

We may assume that each slit in the grating is equivalent to an individual coherent source placed at the mid-point of each slit and sending a wave of amplitude

$$A' = A \, \frac{\sin \alpha}{\alpha} \tag{2.60}$$

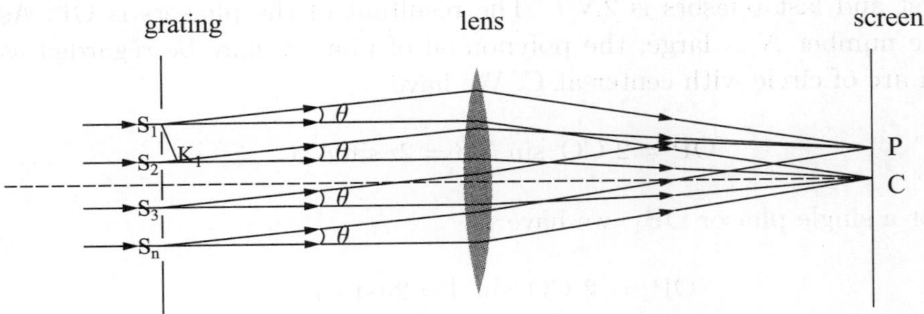

Figure 40: Fraunhoffer diffraction at transmission grating.

at an angle θ with the direction of incident wave. Here, we have

$$\alpha = \frac{\pi}{\lambda}\, b \sin \theta$$

Let S_1K_1 be the perpendicular S_2K_1. Then the path difference between the waves produced from S_1 and S_2 is

$$S_2K_1 = (b + d) \sin \theta$$

and corresponding phase difference is

$$\frac{2\pi}{\lambda}\,(b + d) \sin \theta = 2\beta \quad \text{(say)} \qquad \text{or} \qquad \beta = \frac{\pi}{\lambda}\,(b + d) \sin \theta$$

To find intensity I, we have to superimpose N waves each of amplitude A' differing in phase with near by wave by 2β.

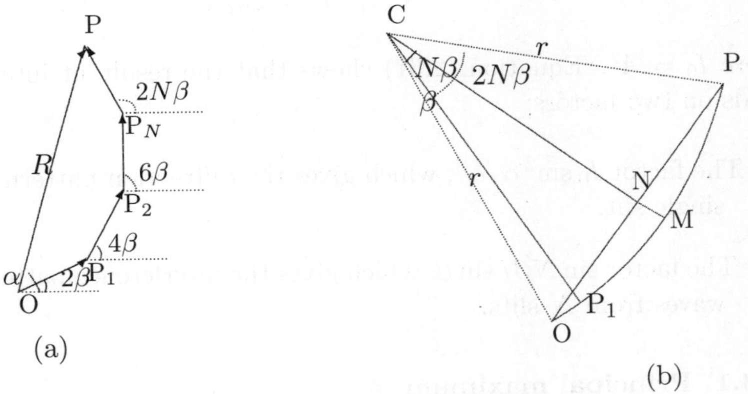

Figure 41: The phasor diagram of N waves.

Figure 41 shows the phasor addition of N phasors. The angle between two successive phasors is 2β. Therefore, the angle between the first and last phasors is $2N\beta$. The resultant of the phasors is OP. As the number N is large, the polynomial of phasors may be regarded as an arc of circle with center at C. We have

$$\text{OP} = 2\,\text{CO}\,\sin N\beta = 2r\sin N\beta$$

For a single phasor OP_1, we have

$$\text{OP}_1 = 2\,\text{CO}\,\sin\beta = 2r\sin\beta$$

Therefore, we have

$$\frac{\text{OP}}{\text{OP}_1} = \frac{\sin N\beta}{\sin\beta}$$

Here, OP is resultant disturbance and OP_1 disturbance of single slit, and therefore, we have

$$A_0 = A'\,\frac{\sin N\beta}{\sin\beta}$$

Using the expression for A' from equation (2.60), we have

$$A_0 = A\,\frac{\sin\alpha}{\alpha}\,\frac{\sin N\beta}{\sin\beta}$$

The resultant intensity is

$$I = A_0^2 = I_0\left(\frac{\sin\alpha}{\alpha}\right)^2\left(\frac{\sin N\beta}{\sin\beta}\right)^2 \tag{2.61}$$

where $I_0 = A^2$. Equation (2.61) shows that the resultant intensity depends on two factors:

(i) The factor $I_0\sin^2\alpha/\alpha^2$, which gives the diffraction pattern due to a single slit.

(ii) The factor $\sin N\beta/\sin\beta$, which gives the interference pattern in the waves from N slits.

12.3.1 Principal maximum

The intensity would be maximum when $\beta = n\pi$. For $\beta = n\pi$, we have

$$\frac{\sin N\beta}{\sin\beta} = \frac{\sin Nn\pi}{\sin n\pi} = \frac{0}{0} \quad \text{(indeterminate)}$$

Therefore, the value can be obtained after differentiation and we have

$$\lim_{\beta \to n\pi} \frac{\sin N\beta}{\sin \beta} = \lim_{\beta \to n\pi} \frac{N \cos N\beta}{\cos \beta} = \frac{N \cos Nn\pi}{\cos n\pi} = \pm N$$

So, the intensity at the central maximum is

$$I = I_0 N^2 \left(\frac{\sin \alpha}{\alpha}\right)^2 \tag{2.62}$$

For $\beta = n\pi$, we have

$$\frac{\pi}{\lambda}(b+d)\sin\theta = n\pi \qquad \text{or} \qquad (b+d)\sin\theta = n\lambda$$

For $n = 0$, we have zeroth order maximum and for $n = \pm 1, \pm 2, \ldots$, we have first, second, $\ldots$ order principal maximum, respectively.

12.3.2 Secondary minima

The intensity is zero when $\sin N\beta = 0$ and $\sin \beta \neq 0$. For $N\beta = m\pi$, we have

$$N\frac{\pi}{\lambda}(b+d)\sin\theta = m\pi \qquad \text{or} \qquad N(b+d)\sin\theta = m\lambda$$

where $m = \pm 1, \pm 2, \ldots, \pm(N-1)$. If $m = 0$ gives principal maximum and $m = N$ also gives principal maximum. The $m = \pm 1, \pm 2, \ldots, \pm(N-1)$ gives minima. There are $(N-1)$ minima between two principal maxima.

12.3.3 Secondary maxima

As there are $(N-1)$ minima between two maxima, there must be $(N-2)$ maxima between two principal maxima. To find out positions of these secondary maxima, we differentiate equation (2.61) with respect to β and equate to zero.

$$\frac{dI}{d\beta} = I_0 \left(\frac{\sin\alpha}{\alpha}\right)^2 \frac{\sin N\beta}{\sin\beta} \left[\frac{N\sin\beta\,\cos N\beta - \sin N\beta\,\cos\beta}{\sin^2\beta}\right] = 0$$

It gives

$$N\sin\beta\,\cos N\beta - \sin N\beta\,\cos\beta = 0 \qquad \text{or} \qquad \tan N\beta = N\tan\beta$$

We have

$$\sec N\beta = \sqrt{1 + \tan^2 N\beta} = \sqrt{1 + N^2\tan^2\beta}$$

It gives

$$\cos N\beta = \frac{1}{\sqrt{1 + N^2 \tan^2 \beta}}$$

Thus, we have

$$\sin N\beta = \tan N\beta \cos N\beta = \frac{N \tan \beta}{\sqrt{1 + N^2 \tan^2 \beta}}$$

Therefore, we have

$$\frac{\sin^2 N\beta}{\sin^2 \beta} = \frac{(N^2 \tan^2 \beta)/(1 + N^2 \tan^2 \beta)}{\sin^2 \beta}$$

$$= \frac{N^2}{\cos^2 \beta + N^2 \sin^2 \beta} = \frac{N^2}{1 + (N^2 - 1) \sin^2 \beta}$$

The intensity at secondary maxima is

$$I = I_0 \left(\frac{\sin \alpha}{\alpha}\right)^2 \frac{N^2}{1 + (N^2 - 1) \sin^2 \beta} \tag{2.63}$$

From equations (2.62) and (2.63), we get

$$\frac{\text{Intensity of secondary maxima}}{\text{Intensity of principal maximum}} = \frac{1}{1 + (N^2 - 1) \sin^2 \beta}$$

When N is large, the intensity of the secondary maxima is very small. Since in the transmission grating, N is very large, the secondary maxima are not visible in the transmission grating.

12.3.4 Angular half-width of principal maxima

The condition for nth principal maximum in the direction θ_n in diffracting grating is

$$(b + d) \sin \theta_n = n\lambda \tag{2.64}$$

Let $(\theta_n + d\theta_n)$ be the direction of the first secondary maximum on the two sides of the nth primary maximum as per positive and negative signs. Then we have

$$(b + d) \sin(\theta_n + d\theta_n) = n\lambda \pm \frac{\lambda}{N} \tag{2.65}$$

On dividing equation (2.65) by (2.64), we get

$$\frac{\sin(\theta_n + d\theta_n)}{\sin \theta_n)} = \frac{n\lambda \pm \lambda/N}{n\lambda}$$

or

$$\frac{\sin\theta_n \, \cos \mathrm{d}\theta_n + \cos\theta_n \, \sin \mathrm{d}\theta_n}{\sin\theta_n} = 1 \pm \frac{1}{nN}$$

For small value of $\mathrm{d}\theta_n$, we have $\sin \mathrm{d}\theta_n = \mathrm{d}\theta_n$ and $\cos \mathrm{d}\theta_n = 1$. Then, we have

$$\frac{\sin\theta_n + \cos\theta_n \, \mathrm{d}\theta_n}{\sin\theta_n} = 1 \pm \frac{1}{nN} \qquad \text{or} \qquad 1 + \frac{\cos\theta_n \, \mathrm{d}\theta_n}{\sin\theta_n} = 1 \pm \frac{1}{nN}$$

It gives

$$\mathrm{d}\theta_n = \frac{1}{nN \cot\theta_n}$$

Here, $\mathrm{d}\theta_n$ relates to half angular width of principal maximum.

12.3.5 Missing orders

The resultant intensity is

$$I = I_0 \left(\frac{\sin\alpha}{\alpha}\right)^2 \left(\frac{\sin N\beta}{\sin\beta}\right)^2 \tag{2.66}$$

where

$$\beta = \frac{\pi}{\lambda}(b+d)\sin\theta \qquad \text{and} \qquad \alpha = \frac{\pi b \sin\theta}{\lambda}$$

The direction of minima for a single slit pattern is expressed as

$$b\sin\theta = m\lambda \tag{2.67}$$

The direction of principal maxima in grating spectra is expressed as

$$(b+d)\sin\theta = n\lambda \tag{2.68}$$

If two conditions expressed by (2.67) and (2.68) are satisfied simultaneously then a particular maximum of nth order is absent in the grating spectrum. These are known as the absent spectra or missing order spectra. On dividing equation (2.68) by (2.67), we get

$$\frac{(b+d)}{b} = \frac{n}{m}$$

This is the condition for the missing order spectra in the grating pattern.

(i) When $d = b$, we have

$$\frac{n}{m} = 2 \qquad \text{or} \qquad n = 2m$$

For $m = 1, 2, 3, \ldots$, we have $n = 2, 4, 6, \ldots$. That is, second, fourth, sixth, $\ldots$ order maxima will be missing.

(ii) When $d = 2b$, we have

$$\frac{n}{m} = 3 \qquad \text{or} \qquad n = 3m$$

For $m = 1, 2, 3, \ldots$, we have $n = 3, 6, 9, \ldots$. That is, third, sixth, ninth $\ldots$ order maxima will be missing.

Exercise 42: Light of wavelength 500 nm falls normally on a plane transmission grating having 5000 line per cm. Find the angle of diffraction for maximum intensity in first order. Given $\sin^{-1} 0.25 = 14°47'$.

Solution: Given $\lambda = 500 \times 10^{-9}$ m $= 5 \times 10^{-5}$ cm, $N = 5000$. We have

$$(b + d) = \frac{1}{N} = \frac{1}{5000}$$

The direction of first maximum is expressed as

$$(b + d) \sin \theta = \lambda \qquad \text{or} \qquad \sin \theta = \frac{\lambda}{(b + d)} = \frac{5 \times 10^{-5}}{1/5000} = 0.25$$

The angle of diffraction is $\theta = \sin^{-1} 0.25 = 14°47'$.

12.4 Resolving power of a plane diffraction grating

Consider a parallel beam of light of wavelengths λ and $(\lambda + d\lambda)$ incident normally on a plane transmission grating having grating element $(b + d)$ and total number of rulings N. The direction of n-th principal maximum for wavelength λ is

$$(b + d) \sin \theta = n\lambda \tag{2.69}$$

The direction of n-th principal maximum for wavelength $(\lambda + d\lambda)$ is

$$(b + d) \sin(\theta + d\theta) = n(\lambda + d\lambda) \tag{2.70}$$

The equation of minima for wavelength λ is

$$N(b + d) \sin \theta = m\lambda \tag{2.71}$$

where m has all the integer values except 0, N, $2N$, $\ldots nN$, as for these values of m, the conditions for the maxima are satisfied. Thus, the first minimum adjacent to n-th principal maximum in the direction $(\theta + d\theta)$ can be obtained by substituting the value of m as $(nN + 1)$ in equation (2.71). Thus, the first minimum in the direction $(\theta + d\theta)$ is

$$N(b + d) \sin(\theta + d\theta) = (nN + 1)\lambda$$

or

$$(b + d)\sin(\theta + \mathrm{d}\theta) = n\lambda + \frac{\lambda}{N} \tag{2.72}$$

From equations (2.70) and (2.72), we get

$$n(\lambda + \mathrm{d}\lambda) = n\lambda + \frac{\lambda}{N} \qquad \text{or} \qquad \frac{\lambda}{\mathrm{d}\lambda} = nN$$

Thus, the resolving power of the grating is nN. This says that the number of rulings per centimeter of a grating should be larger in order to increase its resolving power.

Exercise 43: Light incident normally on a grating of total ruled width of 5.1×10^{-3} m with 2520 lines in all. Fund out if two sodium lines with wavelengths 5890 Å and 5896 Å can be seen distinctly by the grating.

Solution: Given, wavelengths $\lambda_1 = 5890$ Å $= 5.89 \times 10^{-7}$ cm, $\lambda_2 = 5896$ Å $= 5.896 \times 10^{-7}$ cm, $N = 2520$, $n = 1$ and width of ruling 5.1×10^{-3} m. Therefore, we have

$$(b + d) = \frac{5.1 \times 10^{-3}}{2520} = 2.024 \times 10^{-6} \text{ m}$$

For the first order ($n = 1$) of first line, we have

$$\sin \theta_1 = \frac{n\lambda_1}{(b + d)} = \frac{1(5.89 \times 10^{-7})}{2.024 \times 10^{-6}} = 0.2910$$

so that

$$\theta_1 = \sin^{-1}(0.2910) = 16.9178°$$

For the first order ($n = 1$) of second line, we have

$$\sin \theta_2 = \frac{n\lambda_2}{(b + d)} = \frac{1(5.896 \times 10^{-7})}{2.024 \times 10^{-6}} = 0.2913$$

so that

$$\theta_2 = \sin^{-1}(0.2913) = 16.9358°$$

The angular separation between the two lines is

$$\Delta\theta = \theta_2 - \theta_1 = 16.9358 - 16.9178 = 0.018°$$

The resolving power of the grating is

$$\frac{\lambda}{\mathrm{d}\lambda} = \frac{5893 \times 10^{-10}}{6 \times 10^{-10}} = 982$$

As the $nN = 2520$ is more than 982, the two lines of sodium will be seen distinctly by the grating.

12.5 Dispersion power of a plane diffraction grating

We have seen that the light of different wavelengths get dispersed/diffracted by a grating at different angles. In view of this, the angular dispersive power of a diffraction grating is defined as the rate of change of the angle of diffraction with the wavelength of light. It is expressed as

$$D = \frac{d\theta}{d\lambda}$$

For a plane transmission grating, the condition for principal maxima is

$$(b + d)\sin\theta = n\lambda$$

On differentiation, we get

$$(b + d)\cos\theta \; d\theta = n \; d\lambda \qquad \text{or} \qquad \frac{d\theta}{d\lambda} = \frac{n}{(b + d)\cos\theta}$$

Thus, the dispersion power D of a plane diffraction grating is

$$D = \frac{n}{(b + d)\cos\theta}$$

Exercise 44: What is the highest order of spectrum which can be seen with monochromatic light of wavelength 5500 Å by means of a diffraction grating having 4000 lines per cm.

Solution: Given, $\lambda = 5500$ Å $= 5.5 \times 10^{-7}$ m, $N = 4000$ lines per cm. Therefore,

$$(b + d) = \frac{1}{4000} \text{ cm} = 2.5 \times 10^{-6} \text{ m}$$

For the maximum number of order to be seen, we can have $\sin\theta = 1$. Thus, we have

$$n = \frac{(b + d)}{\lambda} = \frac{2.5 \times 10^{-6}}{5.5 \times 10^{-7}} = 4.55$$

The highest order to be seen will be 4.

Exercise 45: How many orders will be observed by a grating having 4260 lines per cm when it is illuminated by a white light having the range from 4000 Å to 7000 Å in wavelength.

Solution: Given,

$$(b + d) = \frac{1}{4260} \text{ cm} = \frac{10^{-2}}{4260} \text{ m}$$

For the maximum number of order to be seen, we can have $\sin\theta = 1$. Thus, we have

$$(b+d)\sin\theta = n\lambda \qquad \text{or} \qquad n = \frac{(b+d)}{\lambda}$$

For the wavelength $\lambda = 4000$ Å $= 4 \times 10^{-7}$ m, we have

$$n = \frac{(b+d)}{\lambda} = \frac{10^{-2}}{4260 \times 4 \times 10^{-7}} = 5.87$$

For the wavelength $\lambda = 7000$ Å $= 7 \times 10^{-7}$ m, we have

$$n = \frac{(b+d)}{\lambda} = \frac{10^{-2}}{4260 \times 7 \times 10^{-7}} = 3.35$$

The maximum number of order to be seen varies from 3 to 5 depending on the wavelength.

13. Electromagnetic radiations

Electromagnetic radiations (waves) of various types are given in Table 1. Between one kind of waves and the next, there are no sharp boundaries. These waves together are known as the electromagnetic spectrum. When we look from γ-rays to radio waves, we find that the wavelength of electromagnetic radiations increases whereas the frequency decreases. It is interesting to note that there is a common property of these waves that they travel with the same velocity c in vacuum. For frequency ν and wavelength λ of a radiation, we have

$$c = \nu\lambda \qquad \text{or} \qquad \nu = \frac{c}{\lambda}$$

Our earth's atmosphere is opaque for most of the part of electromagnetic spectrum. However, the atmosphere is transparent in some frequency bands and two of them are wide enough. These two frequency bands are commonly known as the optical window and radio window.

Table 1. Electromagnetic spectrum

Radiation	Frequency (Hz)	Wavelength (m)
γ-rays	3×10^{18} - 3×10^{22}	10^{-10} - 10^{-14}
x-rays	3×10^{16} - 3×10^{20}	10^{-8} - 10^{-12}
Ultraviolet waves	7.5×10^{14} - 5×10^{17}	4×10^{-7} - 6×10^{-10}
Visible light	4.3×10^{14} - 7.5×10^{14}	7×10^{-7} - 4×10^{-7}
Infrared waves	3×10^{11} - 4.3×10^{14}	10^{-3} - 7×10^{-7}
Microwaves	1×10^{9} - 3×10^{12}	0.3 - 10^{-4}
Radio waves	3×10^{4} - 3×10^{9}	10^{4} - 0.1

13.1 Radio waves

The wavelength of radio waves ranges from more than 10^4 m to about 0.1 m. As the scattering of electromagnetic radiation is inversely proportional to fourth power of its wavelength, the scattering of radio waves is very small and they play very important role in radio astronomy. Radio waves also play important role in the telecommunication with the help of geostationary satellite, as they can transmit through the earth's atmosphere.

Exercise 46: The first molecule in the interstellar medium was identified as OH through its radiation having wavelength 18 cm. Calculate its frequency.

Solution: Given, $\lambda = 18$ cm. The frequency of radiation is

$$\nu = \frac{c}{\lambda} = \frac{3 \times 10^{10}}{18} = 1.67 \times 10^9 \text{ Hz}$$

Exercise 47: In the FM broadcasting, the frequencies used are from 87.5 MHz to 108.0 MHz.

Solution: Given, the frequencies $\nu_1 = 87.5$ MHz $= 8.75 \times 10^7$ Hz and $\nu_2 = 108$ MHz $= 1.08 \times 10^8$ Hz. The corresponding wavelengths λ_1 and λ_2, respectively can be obtained as

$$\lambda_1 = \frac{c}{\nu_1} = \frac{3 \times 10^{10}}{8.75 \times 10^7} = 342.86 \text{ cm}$$

and

$$\lambda_2 = \frac{c}{\nu_2} = \frac{3 \times 10^{10}}{1.08 \times 10^8} = 277.78 \text{ cm}$$

13.2 Microwaves

The wavelength of microwaves range from 0.3 m to 10^{-4} m. Microwaves ovens are an interesting application of microwaves for warming up and cooking of food.

Exercise 48: In a microwave oven, the radiations used have frequencies from 0.3 GHz to 300 GHz. Calculate the wavelength range of radiations.

Solution: Given, the frequencies $\nu_1 = 0.3$ GHz $= 3 \times 10^8$ Hz and $\nu_2 = 300$ GHz $= 3 \times 10^{11}$ Hz. The corresponding wavelengths λ_1 and λ_2, respectively can be obtained as

$$\lambda_1 = \frac{c}{\nu_1} = \frac{3 \times 10^{10}}{3 \times 10^8} = 100 \text{ cm}$$

and

$$\lambda_2 = \frac{c}{\nu_2} = \frac{3 \times 10^{10}}{3 \times 10^{11}} = 0.1 \text{ cm}$$

Thus, the wavelength range of radiations used in microwave ovens is from 0.1 cm to 100 cm.

13.3 Infrared waves

The wavelength of infrared waves range from 10^{-3} m to 7×10^{-7} m. The earth's atmosphere has water (H_2O) and carbon dioxide (CO_2) which absorb the infrared radiations. Consequently, the infrared radiations coming from the interstellar medium do not reach the earth's surface. The infrared radiations are used in the treatment of patients for fomentation.

13.4 Visible light

Since this part of electromagnetic can be seen by our eyes, it is the most familiar form of electromagnetic waves. This part of electromagnetic spectrum is known as the light. The wavelength of visible light range from 7×10^{-7} m to 4×10^{-7} m. The sensitivity of human eye is a function of wavelength. The human eye is most sensitive around 5.5×10^{-7} m.

Exercise 49: Human eye can see the radiation having wavelengths from 4000 Å to 7500 Å. Calculate the range of the visible light in terms of frequencies.

Solution: The frequency ν_1 corresponding to the wavelength $\lambda_1 = 4000$ Å $= 4 \times 10^{-5}$ cm is

$$\nu_1 = \frac{c}{\lambda_1} = \frac{3 \times 10^{10}}{4 \times 10^{-5}} = 7.5 \times 10^{14} \text{ Hz}$$

The frequency ν_2 corresponding to the wavelength $\lambda_2 = 7500$ Å $= 7.5 \times 10^{-5}$ cm is

$$\nu_2 = \frac{c}{\lambda_2} = \frac{3 \times 10^{10}}{7.5 \times 10^{-5}} = 4 \times 10^{14} \text{ Hz}$$

The visible light has a range from 4×10^{14} Hz to 7.5×10^{14} Hz.

13.5 Ultraviolet waves

The wavelength of ultraviolet waves range from 4×10^{-7} m to 6×10^{-10} m. The ultraviolet waves from the outer atmosphere are absorbed by the ozone (O_3) layer available in the earth's atmosphere. Problem of depletion of ozone layer in the earth's atmosphere is being discussed at the international forums. Scientists are worried for the decrease of amount of ozone. The ozone in the earth's atmosphere helps us in protecting us from the ultraviolet radiations coming from the interstellar medium.

13.6 x-rays

The wavelength of x-rays ranges from 10^{-8} m to 10^{-11} m. These rays are produced through transition of electrons in the inner orbits in an atom. These rays have no charge and therefore cannot be deflected by electric or magnetic field. The x-rays with frequencies from 1.5×10^{18} Hz to 3×10^{18} Hz are called as the hard x-rays while those with low frequencies are are called as the soft x-rays. The x-rays are used for radiotherapy of patients. The x-rays are also used in industries.

13.7 γ-rays

The wavelength of γ-rays ranges from 10^{-10} m to 10^{-14} m. These rays are produced in nucleus of an atom. These rays have no charge and therefore cannot be deflected by electric or magnetic field.

14. Multiple choice questions

1. A and B are convex and concave lenses, respectively. Their respective focal lengths are

   (i) positive and positive   (ii) positive and negative

   (iii) negative and positive   (iv) negative and negative

   Ans. (ii)

2. For a lens having surfaces of radii R_1 and R_2, and refractive index μ of material, the focal length is

   (i) $\dfrac{1}{f} = (\mu - 1)\left(\dfrac{1}{R_1} - \dfrac{1}{R_2}\right)$ (ii) $\dfrac{1}{f} = (\mu + 1)\left(\dfrac{1}{R_1} - \dfrac{1}{R_2}\right)$

   (iii) $\dfrac{1}{f} = (\mu - 1)\left(\dfrac{1}{R_1} + \dfrac{1}{R_2}\right)$ (iv) $\dfrac{1}{f} = (\mu + 1)\left(\dfrac{1}{R_1} + \dfrac{1}{R_2}\right)$

   Ans. (i)

3. For a lens having focal length f, the distances of object and its image from the optical point are u and v, respectively. The relation between u , v and f is

(i) $\dfrac{1}{f} = \dfrac{1}{v} - \dfrac{1}{u}$

(ii) $\dfrac{1}{f} = \dfrac{1}{v} + \dfrac{1}{u}$

(iii) $\dfrac{1}{f} = \dfrac{1}{u} - \dfrac{1}{v}$

(iv) $\dfrac{1}{f} = -\dfrac{1}{v} - \dfrac{1}{u}$

Ans. (i)

4. The image of an object is formed by a convex lens between 2F and infinity. The position of the object is

(i) At infinity

(ii) Between infinity and 2F

(iii) Between 2F and F

(iv) Between F and optical point

Ans. (iii)

5. The object is between F and optical point. The image formed by an convex lens is

(i) At infinity

(ii) Between infinity and 2F

(iii) Between 2F and F

(iv) In front of the lens

Ans. (iv)

6. In an achromatic objective of a telescope, we have

(i) one convex and one concave lenses

(ii) both concave lenses

(iii) both convex lenses

(iv) only one convex lens

Ans. (i)

7. Which of the followings is not monochromatic aberration.

(i) Spherical aberration

(ii) Coma

(iii) Astigmatism

(iv) chromatic aberration

Ans. (iv)

8. Two light waves having equal frequency have amplitudes a_1 and a_2, respectively. The intensity of constructive interference is proportional to

(i) $(a_1 - a_2)$ (ii) $(a_1 + a_2)$ (iii) $(a_1 - a_2)^2$ (iv) $(a_1 + a_2)^2$

Ans. (iv)

9. Two light waves having equal frequency have amplitudes a_1 and a_2, respectively. The intensity of destructive interference is proportional to

   (i) $(a_1 - a_2)$ (ii) $(a_1 + a_2)$ (iii) $(a_1 - a_2)^2$ (iv) $(a_1 + a_2)^2$

   Ans. (iii)

10. For a light wave of wavelength λ, the path-difference $\lambda/2$ corresponds to the phase difference

   (i) $\pi/4$ (ii) $\pi/24$ (iii) π (iv) 2π

   Ans. (iii)

11. For constructive interference between two light waves having the same frequency ν and same wavelength λ, the path difference is (m is integer)

   (i) $m\lambda$ (ii) $(m+1/4)\lambda$ (iii) $(m+1/2)\lambda$ (iv) $(m+3/4)\lambda$

   Ans. (i)

12. For destructive interference between two light waves having the same frequency ν and same wavelength λ, the path difference is (m is integer)

   (i) $m\lambda$ (ii) $(m+1/4)\lambda$ (iii) $(m+1/2)\lambda$ (iv) $(m+3/4)\lambda$

   Ans. (iii)

13. For constructive interference between two light waves having the same frequency ν and same wavelength λ, the phase difference is (m is integer)

   (i) $2m\pi$ (ii) $(2m+1/2)\pi$ (iii) $(2m+1)\pi$ (iv) $(2m+3/2)\pi$

   Ans. (i)

14. For destructive interference between two light waves having the same frequency ν and same wavelength λ, the phase difference is (m is integer)

   (i) $2m\pi$ (ii) $(2m+1/2)\pi$ (iii) $(2m+1)\pi$ (iv) $(2m+3/2)\pi$

   Ans. (iii)

15. In the Young's double slit experiment, the separation between two coherent sources of wavelength λ is d. If the screen is placed at a distance D from the sources, the separation between two successive bright fringes is

   (i) $\dfrac{\lambda d}{D}$ (ii) $\dfrac{\lambda D}{d}$ (iii) $\dfrac{d}{\lambda D}$ (iv) $\dfrac{D}{\lambda d}$

Ans. (ii)

16. The diffraction phenomenon is

    (i) bending of light around an obstacle

    (ii) rectilinear propagation of light

    (iii) oscillation of light wave in one direction

    (iv) quantum nature of light Ans. (i)

17. To getting prominent diffraction, the size of the diffracting object should be

    (i) less than the wavelength of light used.

    (ii) of the order of the wavelength of light used.

    (iii) greater than the wavelength of light used.

    (iv) negligibly small. Ans.

18. In Fraunhoffer diffraction, the incident wavefront should be

    (i) elliptical   (ii) spherical   (iii) cylindrical   (iv) plane
    Ans. (iv)

19. Significant diffraction of x-rays can be obtained by

    (i) a single slit   (ii) a double slit

    (iii) an atomic crystal   (iv) a diffraction grating
    Ans. (iii)

20. Which radiation is produced in the nucleus of an atom.

    (i) infrared   (ii) ultraviolet   (iii) x-rays   (iv) γ-rays
    Ans. (iv)

15. Problems and questions

1. Describe about the principal axis, optical point and focal length of a lens.

2. Which of the followings are taken positive or negative.

(a) Distance measured in the direction right to the optical point.

(b) Distance measured in the direction left to the optical point.

(c) Distance measured in upward direction from the principal axis.

(d) Distance measured in downward direction from the principal axis.

Ans.   (a) positive; (b) negative; (c) positive; (d) negative

3. Describe the sign convention used in the optics.

4. What is the spherical aberration? Describe it. How it can be rectified.

5. What is the coma aberration? Describe it. How it can be rectified.

6. What is the astigmatism aberration? Describe it. How it can be rectified.

7. What is the distortion aberration? Describe it. How it can be rectified.

8. What do you understand by diffraction of light?

9. Describe the difference between interference and diffraction.

10. What is a diffraction grating?

11. For diffraction grating, discuss about the intensities of central maximum, secondary minima and secondary maxima.

12. Show that resolving power of a plane diffraction grating increases with the increase of the number of rulings in the grating.

13. Using the expression for intensity I

$$I = I_0 \left(\frac{\sin \alpha}{\alpha}\right)^2 \left(\frac{\sin N\beta}{\sin \beta}\right)^2$$

for diffraction grating having N slits, discuss the reason for absence of secondary maxima.

14. On what factors the dispersion power of a grating depends.

15. Write notes on the following

(i) Sign convention used in optics

(ii) Diffraction grating

(iii) Chromatic aberration

(iv) Spherical aberration

(v) Coma

(vi) Astigmatism

(vii) Distortion

(viii) γ-rays

(ix) x-rays

(x) Ultraviolet radiation

(xi) Infrared radiation

(xii) Microwaves

(xiii) Radio waves

(xiv) Resolving power of a plane diffraction grating

(xv) Dispersion power of a plane diffraction grating

III. Fiber Optics

1. Introduction

For sending information from one place to another, a traditional manner has been to use radio waves (or microwaves) as a carrier. The discovery of laser in 1960 brought a big revolution in the field of telecommunication, as the laser is monochromatic and coherent source of light waves. The frequency of a laser wave is about 10^5 times larger than that of a radio wave. (The wavelength of a laser wave is about 10^5 times smaller than that of a radio wave.) Hence, with the help of the laser waves, about 10^5 times more information can be carried out as compared to that with the help of the radio waves. However, the scattering of an electromagnetic wave is inversely proportional to the fourth power of its wavelength. The scattering of a laser wave is about 10^{20} times larger than that of a radio wave. Consequently, the energy of a laser wave gets dissipated very easily in an open atmosphere. Hence, the laser waves cannot travel long distances in an open atmosphere. Therefore, we require a medium in which there is no loss of energy during propagation from one place to another. In such arrangement, the waves need to be guided also. This requirement for the laser waves has been achieved with the use of an optical fiber. An optical fiber is a very thin glass or plastic conduit designed to guide the light waves along its length. The working of an optical fiber is based on the phenomenon of total internal reflection.

Fiber optics is a technology which uses glass, plastic, or such material for transmission of data. A cable for fiber optics consists of a bundle of fibers which are protected by an outer jacket, made up of treated paper, PVC or metal. Optical fiber has a number of advantages over the copper wire used to make connections electrically. For example, an optical fiber, being made up of glass or plastic, is protected from electromagnetic interference such as that caused by thunderstorms. A single optical fiber has three parts, as shown in Figure 1:

(i) core

(ii) cladding

(iii) sheath (protecting layer)

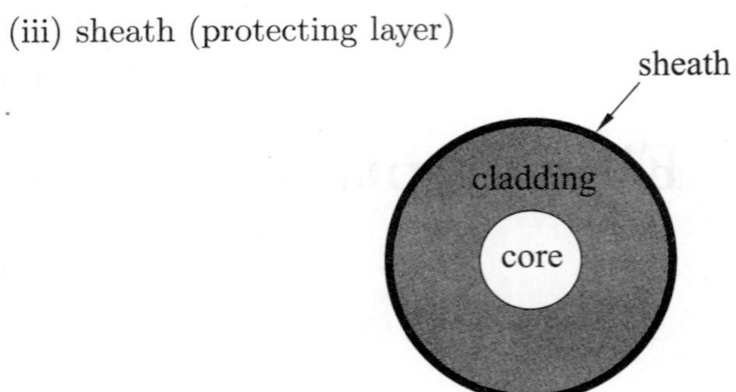

Figure 1: An optical fiber has a core, cladding and sheath.

The shape of a core is cylindrical with small radius. It is situated at the center of the fiber along its axis and is made up of glass. Cladding is a thin cylindrical shell surrounding the core and is made up of an optical material. The refractive index of the material of cladding is smaller than that of the material of core, so that total internal reflection at the interface of the core and cladding could take place when the angle of incidence is larger than the critical angle. Sheath is a plastic coating that protects the fibers from damage and moisture.

In order to understand about the advantages of fiber optics, it is necessary to know about the bandwidth, which is described as the difference between the upper and lower cut-off frequencies of a filter, a communication channel, or a signal spectrum. It is expressed in Hertz. In the case of a low-pass filter or base band signal, the bandwidth is equal to its upper cut-off frequency. In radio communications, bandwidth is the range of frequencies occupied by a modulated carrier wave. For example, an FM radio receiver's tuner spans over a limited range of frequencies. In optics, it is the width of an individual spectral line or the entire spectral range.

Fiber optics has many advantages as compared to the traditional metal communication lines. Some of them are as the following.

(i) Fiber optic cables can carry more data as their bandwidth is greater as compared to a metal cable.

(ii) Fiber optic cables are less susceptible as compared to metal cables to interference.

(iii) Fiber optic cables are much thinner and lighter as compared to metal cables.

(iv) Through fiber optic cables, the data can be transmitted digitally rather than analogically.

(v) Attenuation through fiber optic cables is very low in transmitting the data over a long distance, so there is no need of repeaters.

2. Basic concepts used in optical fiber

In an optical fiber, we employ light waves for carrying digital signals from one place to another. This technology is based on the concept of total internal reflection. The digital signal that is carried by the light is reflected inside the optical cable and hence transfers the information from one place to another. The main concepts of physics that are involved in the optical fibers are the refraction, refractive indices, critical angle and total internal reflection. In the refraction, the light wave bends away from the normal when it propagates from a medium having higher refractive index to a medium having lower refractive index. The phenomena of total internal reflection takes place when the angle of refraction becomes 90°. The incident angle for which the angle of refraction is 90° is known as the critical angle. When a light ray propagating from a higher refractive index medium to a lower refractive index medium has a sufficiently large incident angle, *i.e*, greater than the critical angle, the light gets reflected back into the same medium. For a particular case of an optical fiber whose core is made of glass which is bounced by a plastic cladding the critical angle is 82°. Thus, the light when hits the plastic cladding at an angle more than 82°, then it is reflected back in the same medium, *i.e*, back to the glass core.

3. Total internal reflection

When a ray of light propagates from a denser medium to a rarer one, the angle of refraction is larger than the angle of incidence. If we go on increasing the angle of incidence, at some of its value, the refraction angle becomes equal to 90° and the ray does not enter into the rarer (second) medium, as shown in Figure 2. This phenomenon is known as the total internal reflection and the corresponding angle of incidence is known as the critical angle, denoted by i_c. Thus, we have

$$_1\mu_2 = \mu = \frac{\sin i}{\sin r} = \frac{\sin i_c}{\sin 90} = \sin i_c$$

Thus, the critical angle is

$$i_c = \sin^{-1} \mu$$

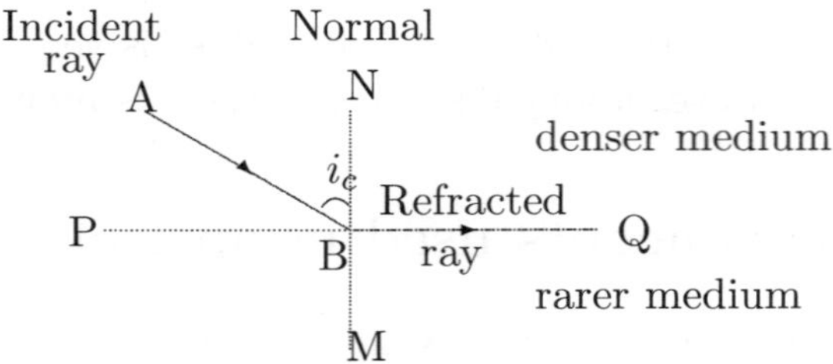

Figure 2: Total Internal Reflection

With the further increase of the angle of incidence, the incident ray is reflected back into the first medium.

Exercise 1: A ray of light propagates from the glass to the air. Calculate the critical angle for total reflection. Refractive index of glass relative to air is 1.5.

Solution: The ray of light goes from glass to air and the refractive index of air relative to glass is

$$_g\mu_a = \frac{1}{_a\mu_g} = \frac{1}{1.5} = 0.6667$$

Thus, the critical angle i_c is

$$i_c = \sin^{-1}\left(_g\mu_a\right) = \sin^{-1}(0.6667) = 41.8°$$

4. Types of optical fibers

Based on the transmission properties and structure, optical fibers are, in general, classified into two categories:

(i) single mode fiber

(ii) multi-mode fiber

As the name implies, a single mode fiber sustains only one mode of propagation, whereas a multi-mode fiber sustains multiple modes. The basic difference from the structural point of view between the two categories is the size of the core. The core diameter of a single mode fiber is about 10 μm and that of a multi-mode fiber ranges from 50 μm to 100 μm, depending on the fiber. When the refractive indices of the core as well as of the cladding are constant, the optical fiber is said to be step-index fiber. On the other side, when the refractive index of the core or/and of the cladding varies, the optical fiber is said to be graded-index fiber.

Figure 3: (a) single-mode fiber; (b) multi-mode fiber.

4.1 Salient features of single mode fibers

(i) For a single mode step index fiber, the core diameter is about $8 - 12$ μm with a cladding thickness of about 125 μm. For a graded index single mode fiber, the core diameter is slightly larger as compared in a step index single mode fiber. This makes handling, connecting and coupling of a graded index fiber a bit easier as compared to a step index fiber.

(ii) For a single mode fiber, the losses are less.

(iii) In a single mode fiber, the cladding is relatively thick. By making the cladding thick, the field at the cladding-air boundary is minimized.

(iv) Both the relative refractive index $\Delta\mu_r$ and numerical aperture N_a, for a single mode fiber are very small. Low N_a means a very small acceptance angle. Thus, the incident ray in a single mode fiber must be nearly perpendicular to the fiber edge.

(v) As only one mode propagates in a single mode fiber, no intermodal dispersion exists there. This makes the single mode fibers suitable for their use with high data rates.

(vi) Manufacturing a single mode fiber is more expensive and difficult. To launch light energy into the fiber, costly laser diodes are required. However, when the high data rate is considered, it becomes cost effective.

(vii) Single mode fibers are becoming more popular for other specialized applications.

4.2 Salient features of multimode fibers

(i) For a multimode step index fiber, the core diameter is about $50 - 250$ μm with a cladding thickness of about $125 - 400$ μm. For a multimode index fiber, the core diameter is about $50 - 100$ μm with a cladding thickness of about $125 - 140$ μm.

(ii) Both the relative refractive index $\Delta\mu_r$ and numerical aperture N_a, for a multimode fiber are large. Thus, the acceptance angle is much larger as compared to that for a single mode fiber. Therefore, the launching of light signal into a multimode fiber is much easier.

(iii) An LED (Light Emitting Device), which is comparatively less expensive, can be used for launching of light into a multimode fiber.

(iv) Handling, connecting and coupling of multimode fibers are easier as compared to those for a single mode fiber.

(v) The major disadvantage in case of multimode fiber is the intermodal dispersion, which limits the data transmission rates.

4.3 Single mode step index fiber

In this class of optical fibers, the refractive indices of the materials of the core (μ_1) and of cladding (μ_2) are both constant, such that $\mu_1 > \mu_2$. When we proceed from the core to the cladding, the refractive index changes in a single step from μ_1 to μ_2. At the interface between the core and cladding, the refractive index changes suddenly in a step, as shown in Figure 4(a). The diameter of the core is about 10 μm whereas the outer diameter of the cladding is about 125 μm, as shown in Figure 4(b). Owing to the small diameter of the core, this fiber allows only one mode to propagate through the core at one time. As compared to a multi-mode fiber, a single mode fiber has a lower signal loss and higher information capacity or bandwidth. These fibers are capable of transferring larger amount of data due to low fiber dispersion. In these fibers, the wavelength can increase or decrease the losses caused by the fiber bending. In general, single mode fibers are considered to be low loss fibers, which increase system bandwidth and length. Therefore, these fibers are most useful for large bandwidth applications. Since these fibers are more resistant to attenuation, they can also be used in significantly longer cable runs.

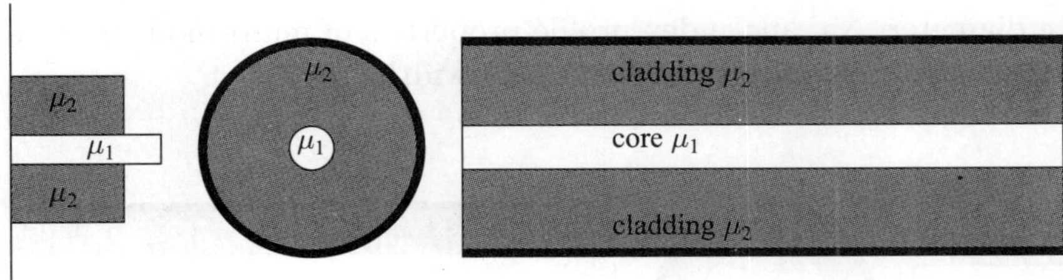

Figure 4: (a) Profile of refractive index of single mode step index fiber; (b) Cross-sectional view of single mode fiber; (c) Longitudinal view of single mode fiber.

4.4 Multi-mode fiber

Contrary to a single-mode fiber, a multi-mode fiber (MMF) allows more than one mode to propagate through the core at one time. For example, over 100 modes can propagate through the core of the fiber at a time. The size of its core is typically about 50 μm whereas the outer diameter of the cladding is about 125 μm, as shown in Figure 3(b). The multi-mode fibers are of two categories:

(i) Multi-mode step index fiber

(ii) Multi-mode graded index fiber

(i) Multi-mode step index fiber

In this class the refractive indices of the materials of the core μ_1 and cladding μ_2 are both constant. At the interface between the core and cladding, the refractive index changes suddenly in a step, as shown in Figure 5 (a). In this type of optical fiber, the number of propagating modes depends on the ratio of core diameter to the wavelength. This ratio is inversely proportional to the numerical aperture (abbreviated as Na and discussed later on). Typically the core diameter is 50 μm to 100 μm and Na varies from 0.20 to 0.29, respectively. Multi-mode fiber is used in short lengths, such as those used in Local Area Networks (LANs) and Storage Area Networks (SLANs). Because the multi-mode optical fiber has larger Na and the larger core size, fiber connections and launching of light is very easy. Multi-mode fibers permit the use of light emitting diodes (LEDs). In such fibers, core-to-core alignment is less critical during fiber spacing. However, due to several modes, the effect of dispersion gets increased, *i.e.*, the modes arrive at the fiber end at slightly different times and therefore spreading of pulses takes place.

This dispersion of the modes affects the system bandwidth. Hence, the core diameter, Na, and index profile properties of multi-mode fibers are optimized to maximize the system bandwidth.

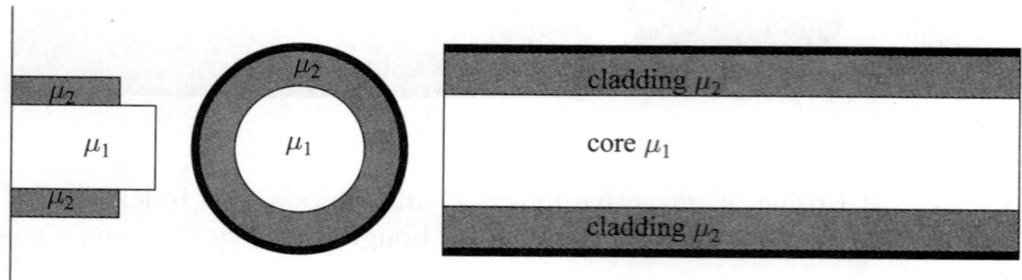

Figure 5: (a) Profile of refractive index of multi-mode index fiber; (b) Cross-sectional view of multi-mode fiber; (c) Longitudinal view of multi-mode fiber.

(ii) Multi-mode graded index fiber

In a multi-mode graded index fiber, the refractive index of the core decreases with increasing radial distance from the fiber axis, which is the imaginary central axis running along the length of the fiber, as shown in Figure 6. The value of the refractive index is the highest at the center of the core and decreases to a value at the edge of the core that equals the refractive index of the cladding. Hence, the light rays in the outer zones of the core travel faster than those in the center of the core. Hence, the dispersion of the modes is compensated by this type of fiber design. Under this situation, the light rays follow sinusoidal paths along the fiber. In such a fibers, the most common profile of the refractive index is very nearly parabolic that results in continual refocusing of the rays in the core, and minimizing modal dispersion. Standard graded index fibers typically have a core diameter of 50 μm or 62.5 μm and a cladding diameter of 125 μm. It is typically used for transmitting the information to the distance to a couple of kilometers. The advantage of the multi-mode graded index fiber in comparison with multi-mode step index fiber is the considerable decrease in the modal dispersion.

5. Parameters of fiber optics

As mentioned earlier, the main phenomenon used in the fiber optics technique is the total internal reflection at the interface between the core and cladding in a fiber cable. For total internal reflection, the light must travel from a medium of large refractive index to another medium

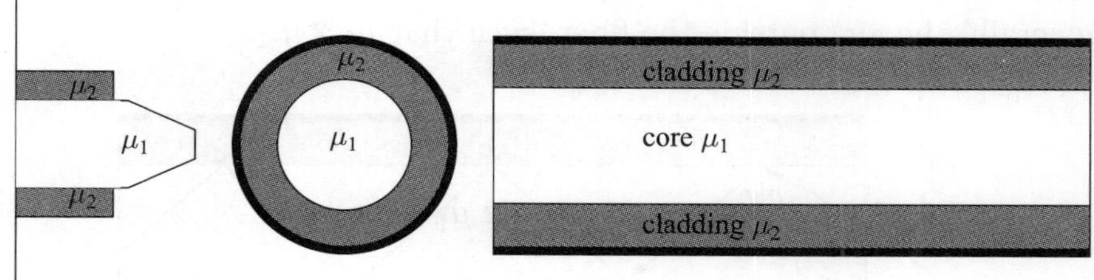

Figure 6: (a) Profile of refractive index of multi-mode graded index fiber; (b) Cross-sectional view of multi-mode fiber; (c) Longitudinal view of multi-mode.

of low refractive index and the angle of incidence must be larger than the critical angle. Suppose the refractive indices of the materials of the core and cladding are μ_1 and μ_2, respectively, such that $\mu_1 > \mu_2$. Some parameters pertaining to the fiber optics are as the following.

(i) Relative refractive index

(ii) Acceptance angle

(iii) Acceptance cone

(iv) Numérical aperture

(v) Skip distance

5.1 Relative refractive index

When the refractive indices of the materials of the core and cladding in an optical fiber are μ_1 and μ_2, respectively, the relative refractive index $\Delta\mu_r$ is expressed as

$$\Delta\mu_r = \frac{\mu_1 - \mu_2}{\mu_1} = 1 - {}_1\mu_2$$

where ${}_1\mu_2$ is the refractive index of cladding relative to that of in core. It is also known as the fractional refractive index. Obviously, it is a measure of the difference between the refractive indices of the materials of the core and cladding. Larger the difference, larger the relative refractive index.

5.2 Acceptance angle

Let us consider a cylindrical optical fiber (Figure 7) where μ_1 and μ_2 are refractive indices of the materials of the core and cladding, respectively, such that $\mu_1 > \mu_2$. Suppose μ_0 is refractive index of medium (which is generally the air) outside the fiber, such that $\mu_0 < \mu_1$.

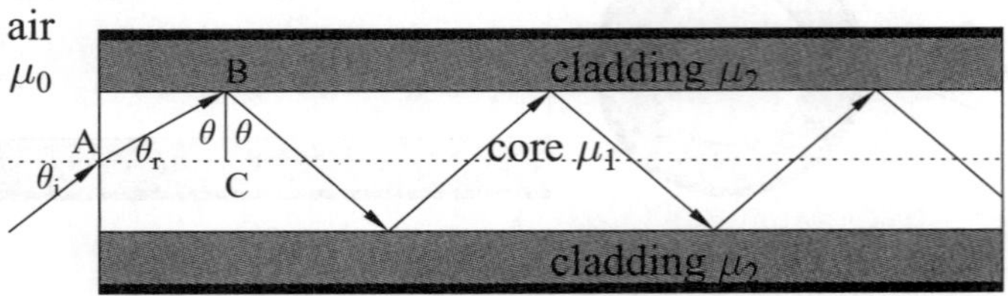

Figure 7: At one end of an optical fiber, the light incident at an angle θ_i.

At one end of the optical fiber, suppose the light incident at an angle θ_i. The angle of refraction is θ_r such that $\theta_r < \theta_i$. Then, the light travels in the core and incident at the core-cladding interface at an angle θ so that $\theta_r + \theta = 90°$. When the angle θ is larger than the critical angle θ_c, the light undergoes the total internal reflection at the interface between the core and cladding. As long as the angle θ is larger than the critical angle θ_c, the light will be reflected back into the core. Let us now compute the incident angle θ_i for which $\theta \geq \theta_c$. Following the Snell's law, we have

$$\mu_0 \sin \theta_i = \mu_1 \sin \theta_r \qquad \text{or} \qquad \frac{\sin \theta_i}{\sin \theta_r} = \frac{\mu_1}{\mu_0} \qquad (3.1)$$

With the increase of the angle θ_i, the angle θ_r increases and consequently, the angle θ decreases. For the fiber optics technique, obviously, we can not afford to increase the value of θ_i beyond a certain limit denoted by θ_m. This value θ_m of the incidence angle is known as the acceptance angle. Corresponding to θ_m, we have $\theta = \theta_c$. From the Δ ABC, it is seen that

$$\sin \theta_r = \sin(90° - \theta) = \cos \theta \qquad (3.2)$$

From equations (3.1) and (3.2), we have

$$\frac{\sin \theta_i}{\cos \theta} = \frac{\mu_1}{\mu_0} \qquad \text{or} \qquad \sin \theta_i = \frac{\mu_1}{\mu_0} \cos \theta$$

When $\theta = \theta_c$, we have $\theta_i = \theta_m$ and therefore

$$\sin \theta_m = \frac{\mu_1}{\mu_0} \cos \theta_c \qquad (3.3)$$

At the critical angle θ_c, from the Snell's law, we have

$$\mu_1 \sin \theta_c = \mu_2 \sin 90° \qquad \text{or} \qquad \sin \theta_c = \frac{\mu_2}{\mu_1}$$

We have

$$\cos \theta_c = \sqrt{1 - \sin^2 \theta_c} = \sqrt{\frac{\mu_1^2 - \mu_2^2}{\mu_1^2}} \tag{3.4}$$

Using equation (3.4) in (3.3), we have

$$\sin \theta_m = \frac{\sqrt{\mu_1^2 - \mu_2^2}}{\mu_0} \tag{3.5}$$

When the incident ray of light is launched from air medium (for which $\mu_0 = 1$), we have

$$\sin \theta_m = \sqrt{\mu_1^2 - \mu_2^2} \qquad \text{or} \qquad \theta_m = \sin^{-1}(\sqrt{\mu_1^2 - \mu_2^2})$$

The angle θ_m is known as the acceptance angle for the optical fiber. The value of θ_m is always less than $90°$. That is when the incidence angle θ_i is less than θ_m, the light at the interface between the core and cladding will show the phenomenon of total internal reflection.

5.3 Acceptance cone

An optical fiber is cylindrical in shape. Obviously, the light can incident from any direction around the axis of the fiber. Thus, all these directions form a cone, as shown in Figure 8. When the incidence angle θ_i is equal to the acceptance angle θ_m, the corresponding cone is known as the acceptance cone.

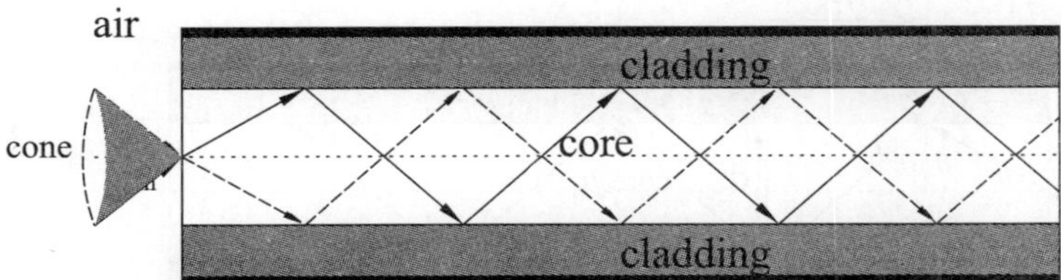

Figure 8: Acceptance cone.

Thus, a light ray lying in the acceptance cone will show the phenomenon of total internal reflection. Obviously, a light ray lying outside the acceptance cone will be refracted into the cladding and the energy will be lost.

5.4 Numerical aperture

Numerical aperture N_a is another parameter of an optical fiber. It is a measure of light gathering ability of the optical fiber and is expressed as

$$N_a = \mu_0 \sin \theta_m$$

where μ_0 is the refractive index of the medium outside the optical fiber and θ_m the acceptance angle. It is a dimensionless quantity. When the medium outside the fiber is air ($\mu_0 = 1$), the value of N_a is less than one, as the value of $\sin\theta_m$ is always less than one. Using the value of $\sin\theta_m$, we get

$$N_a = \sqrt{\mu_1^2 - \mu_2^2}$$

This relation shows that the light gathering ability of an optical fiber increases with its numerical aperture. As the value of $\sin\theta_m$ is always less than 1, the value of the numerical aperture is less than 1. The value of N_a varies in the range from 0.14 to 0.5.

5.5 Skip distance

As mentioned earlier, the propagation of light in an optical fiber is based on the phenomenon of total internal reflection. A light ray gets reflected at the interface between the core and cladding. The distance along the length of the fiber, between two successive reflections of a ray of light propagating in the fiber is known as the skip distance L_s. In Figure 8, the distance BD is the skip distance and is expressed as

$$L_s = d\cot\theta_r \tag{3.6}$$

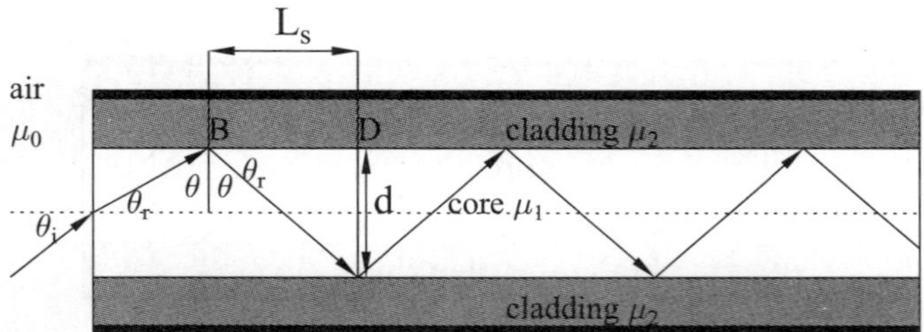

Figure 9: Skip distance.

where d is the diameter of the core and θ_r the angle of refraction in the core. When μ_0 and μ_1 are refractive indices of out side medium and of core, respectively, the Snell's law is

$$\frac{\sin\theta_i}{\sin\theta_r} = \frac{\mu_1}{\mu_0} \qquad \text{or} \qquad \sin\theta_r = \frac{\mu_0}{\mu_1}\sin\theta_i$$

Thus, we have

$$\cos\theta_r = \sqrt{1 - \left(\frac{\mu_0\sin\theta_i}{\mu_1}\right)^2}$$

and

$$\cot \theta_r = \frac{\cos \theta_r}{\sin \theta_r} = \sqrt{\left(\frac{\mu_1}{\mu_0 \, \sin \theta_i}\right)^2 - 1} \qquad (3.7)$$

Using equations (3.7) in (3.6), we have

$$L_s = d \sqrt{\left(\frac{\mu_1}{\mu_0 \, \sin \theta_i}\right)^2 - 1}$$

It is obvious that the inverse of the skip distance, *i.e.*, $1/L_s$ will give the total number of reflections made by the light ray per unit length of the fiber. For an fiber of length L, the number of reflections N_r will be

$$N_r = \frac{L}{L_s} = \frac{L}{d\sqrt{(\mu_1/\mu_0 \, \sin \theta_i)^2 - 1}}$$

Exercise 2: The refractive indices of the materials of core and cladding of a step index optical fiber are 1.52 and 1.40, respectively. Calculate (i) critical angle, (ii) numerical aperture, (iii) acceptance angle, and (iv) fractional refractive index.

Solution: Given, refractive index of the material of core $\mu_1 = 1.52$ and that of the cladding $\mu_2 = 1.40$. (i) The critical angle is

$$\theta_c = \sin^{-1}(\mu_2/\mu_1) = \sin^{-1}(1.40/1.52)$$

$$= \sin^{-1}(0.9211) = 67.09°$$

(ii) The numerical aperture is

$$N_a = \sqrt{\mu_1^2 - \mu_2^2} = \sqrt{(1.52)^2 - (1.40)^2} = 0.5919$$

(iii) The acceptance angle is

$$\theta_0 = \sin^{-1}\sqrt{(\mu_1^2 - \mu_2^2)} = \sin^{-1}\sqrt{(1.52)^2 - (1.40)^2}$$

$$= \sin^{-1}(0.5919) = 36.29°$$

(iv) The fractional refractive index is

$$\Delta \mu_r = \frac{\mu_1 - \mu_2}{\mu_1} = \frac{1.52 - 1.40}{1.52} = 0.0789$$

Exercise 3: A light ray enters from air to a optical fiber. The refractive index of air is 1.0. The refractive indices of the materials of core and cladding of a step index optical fiber are 1.51 and 1.46, respectively. Calculate (i) critical angle, (ii) numerical aperture, (iii) acceptance angle, and (iv) fractional refractive index.

Solution: Given, refractive index of air $\mu_0 = 1$, that of the material of the core $\mu_1 = 1.51$ and that of the material of the cladding $\mu_2 = 1.46$.
(i) The critical angle is

$$\theta_c = \sin^{-1}(\mu_2/\mu_1) = \sin^{-1}(1.46/1.51)$$

$$= \sin^{-1}(0.9669) = 75.21°$$

(ii) The numerical aperture is

$$N_a = \sqrt{\mu_1^2 - \mu_2^2} = \sqrt{(1.51)^2 - (1.46)^2} = 0.3854$$

(iii) The acceptance angle is

$$\theta_0 = \sin^{-1}\sqrt{(\mu_1^2 - \mu_2^2)} = \sin^{-1}\sqrt{(1.51)^2 - (1.46)^2}$$

$$= \sin^{-1}(0.3854) = 22.67°$$

(iv) The fractional refractive index is

$$\Delta\mu_r = \frac{\mu_1 - \mu_2}{\mu_1} = \frac{1.51 - 1.46}{1.51} = 0.0331$$

Exercise 4: A glass clad fiber is made with refractive index 1.5 and the cladding is doped to give a fractional refractive index of 0.0005. Determine (i) the cladding refractive index, (ii) the critical refraction angle, (iii) the critical acceptance angle, and (iv) the numerical aperture.

Solution: We have refractive index of core $\mu_1 = 1.5$ and the fractional refractive index is

$$\frac{\mu_1 - \mu_2}{\mu_1} = 0.0005$$

Using $\mu_1 = 1.5$, we have

$$\frac{1.5 - \mu_2}{1.5} = 0.0005 \qquad \text{or} \qquad \mu_2 = 1.5 - 1.5 \times 0.0005 = 1.49925$$

When θ_r is the angle of refraction and light enters the core from outside, for total internal reflection we have

$$\mu_1 \sin(90 - \theta_r) = \mu_2 \sin 90 \qquad \text{or} \qquad \cos \theta_r = \frac{1.49925}{1.5} = 0.9995$$

It gives $\theta_r = 1.81°$. Thus, the angle for critical refraction is $90 - 1.81 = 88.19°$. For the incidence angle θ_i, we have

$$\mu_0 \sin \theta_i = \mu_1 \sin \theta_r \qquad \text{or} \qquad \sin \theta_i = \frac{1.5}{1.0} \sqrt{1 - \cos^2 \theta_r}$$

Thus, we have

$$\sin \theta_i = 1.5\sqrt{1 - (0.9995)^2} = 0.04743 \qquad \text{or} \qquad \theta_i = 2.72°$$

Hence, the critical acceptance angle is $2.72°$. The numerical aperture is

$$N_a = \sqrt{\mu_1^2 - \mu_2^2} = \sqrt{(1.5)^2 - (1.49925)^2} = 2.24 \times 10^{-3}$$

Exercise 5: The refractive indices of the materials of the core and cladding of a step index optical fiber of diameter 0.062 mm are 1.54 and 1.38, respectively. Calculate (i) critical angle, (ii) numerical aperture, (iii) acceptance angle, (iv) fractional refractive index, (v) number of reflections in 80 cm of fiber for a ray at the maximum incidence angle.

Solution: Given, $d = 0.062$ mm, $L = 80$ cm, refractive index of air $\mu_0 = 1$, that of the material of the core $\mu_1 = 1.54$ and that of the material of the cladding $\mu_2 = 1.38$. (i) The critical angle is

$$\theta_c = \sin^{-1}(\mu_2/\mu_1) = \sin^{-1}(1.38/1.54) = \sin^{-1}(0.8961) = 63.65°$$

(ii) The numerical aperture is

$$N_a = \sqrt{\mu_1^2 - \mu_2^2} = \sqrt{(1.54)^2 - (1.38)^2} = 0.6835$$

(iii) The acceptance angle is

$$\theta_0 = \sin^{-1}\sqrt{(\mu_1^2 - \mu_2^2)} = \sin^{-1}\sqrt{(1.54)^2 - (1.38)^2}$$

$$= \sin^{-1}(0.6835) = 43.12°$$

(iv) The fractional refractive index is

$$\Delta\mu_r = \frac{\mu_1 - \mu_2}{\mu_1} = \frac{1.54 - 1.38}{1.54} = 0.1039$$

(v) The number of reflections

$$N_r = \frac{L}{d\sqrt{(\mu_1/\mu_0 \, \sin\theta_i)^2 - 1}}$$

$$= \frac{80 \times 10^{-2}}{0.062 \times 10^{-3}\sqrt{(1.54/1.0 \times 0.6835)^2 - 1}} = 6390.79$$

The number of reflections is 6390.

6. V-number of a fiber

The V-number of a fiber is defined as

$$V = \frac{2\pi a}{\lambda} \, (\mu_1^2 - \mu_2^2)^{1/2}$$

where a is the radius of fiber core and λ the operating wavelength. The V-number is dimensionless and often referred to as the normalized frequency. The parameter V is related to the number of modes a fiber can support. Detailed calculations give the number of modes M supported in a multimode step index fiber as

$$M_{SI} = \frac{1}{2}\frac{(2\pi a)^2}{\lambda^2} \, (\mu_1^2 - \mu_2^2) = \frac{V^2}{2}$$

When two possible polarizations are accounted for, the number of modes gets

$$M_{SI} = 2 \times \frac{V^2}{2} = V^2$$

The number of modes in a graded index fiber is about half of that in a step index fiber. Thus, the number of modes in a graded index fiber is

$$M_{GI} = \frac{M_{SI}}{2} = \frac{V^2/2}{2} = \frac{V^2}{4}$$

On accounting for two possible polarizations, we have

$$M_{GI} = 2 \times \frac{V^2}{4} = \frac{V^2}{2}$$

7. Fiber optics communication

A block diagram of a communication system often used before the development of optical fiber communication system is shown in Figure 10. The essential components of the system are a modulator or transmitter, transmission medium and the demodulator or receiver.

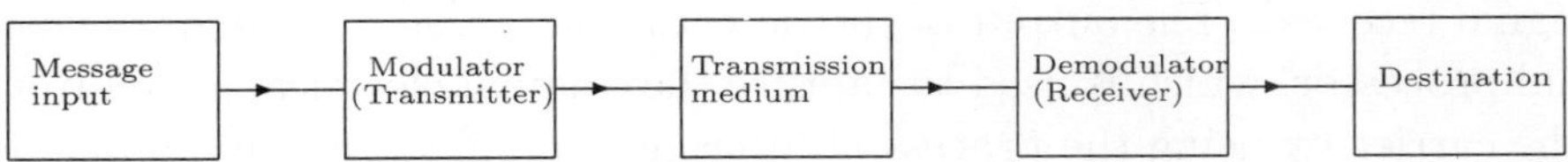

Figure 10: Block diagram of a communication system used in the early days.

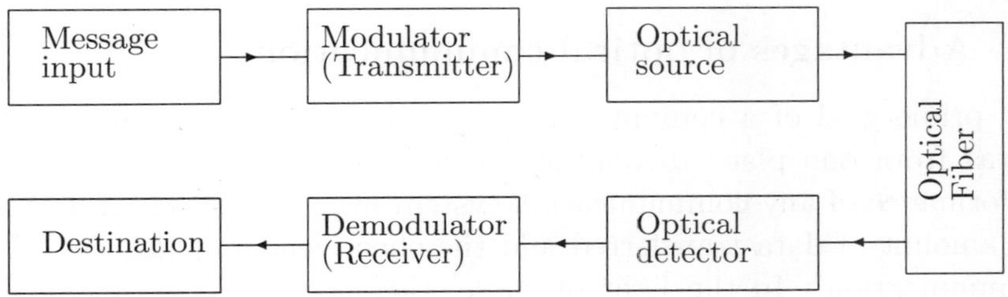

Figure 11: Block diagram of fiber optics communication system.

The optical fibers have replaced copper coaxial cables due to their various advantages as they are very light weight, thin conduit cables which provide large communication capacity with lower loss. A block diagram of a fiber optic communication system from signal source to signal output is shown in Figure 11. Here, the information that is to be transmitted is first converted into an optical signal from an electrical signal. Then the optical signal is further converted into an electrical after transmission by an optical fiber. Independent of the original nature of the signal, a fiber provides the choice format of transmission as analog and digital because these two formats are convertible from one to other. Therefore, the signal in analog or digital form is impressed into the carrier wave with the help of a modular. The carrier wave is generated from the carrier source which may be either light emitting diode (LED) or laser diode (LD). This carrier wave is modulated using various techniques, *viz.* frequently modulation, amplitude modulation and digital modulation. The carrier source output into the optical fiber is represented by a single pulse. When a pulse is passed through a fiber, then it is attenuated and distorted due to several mechanisms, for example, by inter-modal distortion. Therefore, repeaters and regenerators are used to amplify the light signal at several positions of the fiber. And after that the light signal at several positions of the fiber. And after that the light is coupled into a detector that may be a semiconductor device or most commonly used a PIN diode at the end of a fiber. This changes the optical signal back into an electrical signal. The response of a detector should be well matched with the optical frequency of the signal received. The output of the detector then passes through the signal processor, which is used to capture the original electrical signal from the carrier by using the process of filtering, amplification and an analog to digital conversion. The signal output is finally communicated by the cathode ray tube (when it is a video signal), by the loudspeaker (when it is a audio signal) or by the computer input (when it is a digital signal).

7.1 Advantages of optical communication

The prime goal of a communication system is to transfer information (data) from one place to another in an effective manner. Thus, the performance of any communication system may be assessed in terms of the amount of data transferred and the effectiveness of the way of the communication. In the light of these, the merits of the optical fiber communication over the conventional (radio and microwave) methods may be considered as the following.

(i) **High capacity for carrying information**: As we know the light waves have extremely high capacity for carrying information from one place to another. Thus, with the invent of optical fibers, the bandwidth available is extremely large.

(ii) **Small size and small weight**: Optical fibers are so thin that their diameter is less than that of human hair. Thus, even along with the protective coating, the optical fibers are much lighter than copper cables used in the conventional communication. Hence, the optical fibers occupy very little space in the already crowded ducts in the cities.

(iii) **Immunity to interference**: As an optical fiber is a dielectric waveguide, it has no effect of electromagnetic pulses etc and can be in an environment where electrical noise is present. Thus, the fiber cables require no shielding from electromagnetic interference. It is free from lightening strikes when used overhead.

(iv) **Negligible cross talk**: As the fields are confined to a fiber, the optical interference between fibers is negligible. The cross talk is negligible even when a number of fibers are cabled together.

(v) **Low transmission loss**: The optical fibers have been developed with losses as low as 0.2 dB/km or less. It allows the implementation of communication links with extremely large repeater spacing.

(vi) **Ruggedness and flexibility**: Because of protective coatings and proper cabling, it is possible to construct optical fibers having high tensile strength. In terms of storage, transportation handling, and installation, optical cables are generally superior to copper cables.

(vii) **Signal security**: As the light from optical fibers do not radiate significantly, and therefore it is difficult to tap information during its transmission.

(viii) **System reliability**: The reliability of optical components is considered from 20 to 30 years. This value of reliability is much better than that with the conventional electrical conductor systems.

(ix) **Low cost**: The cost of semiconductor lasers, photodiodes, connectors, and couplers are on the higher side. The number of repeater stations required is less. Taking all the factors such as the information carrying capacity, system reliability, predicted lifetime, and security, the overall system cost is very low.

Owing to aforesaid advantages, light wave communication is becoming rapidly an important mode of transmission of information from one place to another.

7.2 High bit rate fiber communication

A high bit rate signal carried on copper wire transmission line is generally needed to be amplified after every 300 m. However, high bit rate signals when carried on an optical fiber, the need of such amplification is required after every 100 km or so. As discussed earlier, a detector changes the optical signal back into an electrical signal, the light signal is coupled to the detector at the remote end of the fiber. This is done effectively when the response of the detector is well matched with the optical frequency of the signal received. Then a signal processor handles the detector output. The function of the signal processor is to recapture the original signal from the carrier. The process involves filtration and amplification, and a digital to analog conversion.

7.3   Allowed modes of normalized frequency

It appears from the theory of acceptance cone that every ray shall propagate successfully once it enters the fiber within its acceptance cone. However, this in not the case always and only certain ray directions or modes are allowed to propagate successfully. Actually a ray represents plane waves which move up and down in the fiber. Evidently, such waves overlap and interfere with one another and only those waves will sustain which satisfy the condition of resonance. Considering this point, we can derive a relation for a parameter m_m in terms of the core diameter d, numerical aperture N_a and the wavelength λ as

$$m_m = \frac{1}{2}\left(\frac{\pi d N_a}{\lambda}\right)^2$$

The largest integer that is less than the parameter m_m shall give the maximum number of modes which can propagate successfully through the fiber. Therefore, it is clear that the number of possible modes will be larger for the higher value of the ratio d/λ. So, a fiber of big diameter allows large number of modes to propagate through it. For this reason, they are known as the multi-mode fiber. However, when d/λ is small such that m_m is less than 2, the fiber will allow only one mode to propagate through it. Such type of fiber is therefore known as the single mode fiber or mono-mode fiber. The condition $m_m < 2$ for a single mode fiber can be achieved when

$$\frac{d}{\lambda} < \frac{2}{\pi N_a}$$

The above condition, related to the diameter, guarantees the performance of single mode fiber. However, a more careful analysis shows that the single mode performance can be achieved even when

$$\frac{d}{\lambda} < \frac{2.4}{\pi N_a}$$

It is evident that the parameter m_m decides the number of possible modes. Since this parameter depends on core diameter d and the numerical aperture N_a, the number of allowed modes would be different for fibers of different core diameters. the word 'number' intuitively adds a concept of normalized frequency, expressed as

$$\nu_n = \frac{\pi d N_a}{\lambda} = \frac{\pi d}{\lambda}\sqrt{\mu_1^2 - \mu_2^2}$$

It shows that the normalized frequency is nothing but the factor carried by the parentheses of the parameter m_m. Thus, in terms of the normalized frequency ν_n, we have

$$m_m = \frac{\nu_n^2}{2}$$

Exercise 6: A graded optical fiber has a core diameter of 0.06 mm and numerical aperture of 0.23 at a wavelength 8600 Å. Calculate (i) the normalized frequency, and (ii) number of modes guided in the core.

Solution: Given, $d = 0.06$ mm, numerical aperture $N_a = 0.23$, and wavelength $\lambda = 8600$ Å. (i) The normalized frequency

$$\nu_n = \frac{\pi d N_a}{\lambda} = \frac{3.14 \times 0.06 \times 10^{-3} \times 0.23}{8600 \times 10^{-10}} = 50.39$$

(ii) The number of modes guided in the core.

$$m_m = \frac{1}{2} \nu_n^2 = \frac{1}{2} (50.39)^2 = 1269.58$$

The number of modes guided in the core is 1269.

8. Fiber losses

The following three losses may occur in an optical fiber.

(i) Absorption

(ii) Geometric effects

(iii) Rayleigh scattering

8.1 Absorption

Even a pure glass absorbs light of specific wavelength. Strong electronic absorption occurs in the UV region and vibrational absorption occurs in the IR region of wavelengths from 7 μm to 12 μm. These losses are attributed due to inherent property of glass and is known as the intrinsic absorption. Further, this loss is insignificant.

Impurities are major extrinsic sources of losses in an optical fiber. Hydroxyl radical OH and transitions metals like Ni, Cr, Cu, Mn etc. have electronic absorption near visible range of spectrum. These impurities should be kept away, as far as possible, from the fiber. Intrinsic as well as extrinsic losses are found minimum around 1.3 μm.

8.2 Geometric effects

These losses may occur due to manufacturing defects like irregularities in fiber dimensions during drawing process or during coating, cabling or insulation processes.

8.3 Rayleigh scattering

As glass has disordered structure having local microscopic variation in density which may also cause variation in the refractive indices. Thus, the light propagating through these structures may suffer scattering losses due to Rayleigh scattering, which is inversely proportional to the fourth power of wavelength of radiation. It shows that Rayleigh scattering sets a lower limit on wavelength that can be transmitted by a glass fiber at 0.8 μm below for which the scattering loss is appreciably high.

9. Dispersion

We know that a pulse launched with a fiber gets attenuated due to the losses in a fiber. Further, the incoming pulse also spreads during the transit through the fiber. Therefore, a pulse at the output is wider as compared to that at the input. Hence, the pulse gets distorted as it moves through the fiber. This distortion of pulse is due to the dispersion effects which is measured in terms of nanoseconds per km.

There are three phenomena that may contribute towards the distortion effects as the following.

(i) Material dispersion

(ii) Wave guide dispersion

(iii) Intermodal dispersion

9.1 Material dispersion

Material dispersion occurs as the refractive index of material varies with the wavelength of radiation. Since the group velocity V_g of a mode is a function of the refractive index, various spectral components of a given mode will travel at different speeds, depending on the wavelength. Material dispersion is therefore an intramodal dispersion effect. For the refractive index $\mu(\lambda)$, the material dispersion is expressed as

$$D_{mat} = \lambda \, \frac{\mathrm{d}^2 \mu}{\mathrm{d}\lambda^2} \tag{3.8}$$

9.2 Wave-guide dispersion

The effect of waveguide dispersion on pulse spreading can be approximated by assuming that the refractive index of the material is independent of wavelength

Wave-guide dispersion arises due to the guiding properties of fiber. Amount of wave-guide dispersion may also be expressed in a manner similar to equation (3.8) where the material refractive index is replaced by the effective refractive index. The effective refractive index for any mode of propagation varies with wavelength which may cause pulse spreading in the same manner as the variation of refractive index in case of material dispersion.

9.3 Intermodal dispersion

A ray of light passing through a fiber follows a zig-zag path and when a number of modes are moving through a fiber, they will move with different net velocities with respect to the fiber axis. It shows that some modes will arrive at the output earlier than the others. That is, there is spread of the input pulse. This process is known as the intermodal dispersion. It may be noted that this kind of dispersion does not depend on the spectral width of the source. That is, a light pulse from a pure monochromatic source will still be giving intermodal dispersion.

In case of MMF, all the three spreading mechanisms are observed simultaneously whereas in case of SMF, only material and wave guide dispersion are observable.

In the case of fibers with low numerical aperture, smaller dispersion is observed while for a fiber with high numerical aperture, large dispersion is observed. Dispersion may be restricted by the use of a low numerical aperture and a narrow spread width source.

10. Transmission windows for optical fibers

The wavelength range in which the fiber communication takes place is from 0.7 to 1.7 μm. As we stated earlier, the major sources responsible for the loss of light in a fiber during transmission are: (i) impurities in the material, (ii) scattering of light, and (iii) geometry of the transmission line. Taking into account all kinds of possible losses, there are some regions where the net loss is small. These regions are used for the transmission purpose and are known as the transmission windows. The ranges of these windows are as the following.

(i) from 0.80 to 0.90 μm region, called the first window

(ii) from 1.26 to 1.34 μm region, called the second window

(iii) from 1.50 to 1.63 μm region, called the third window

Suppose P_i is the input power to a fiber and P_o the output power at a distance z. The fiber attenuation (loss), denoted by α is expressed as

$$\alpha = \frac{10}{z} \log\left(\frac{P_i}{P_0}\right)$$

Generally, a low loss fiber may have average loss of 1 dB/km at 900 nm. In the present days, commonly used fibers have losses as low as 0.1 dB/km.

11. Multiple choice questions

1. Carrier waves used in the optical fiber communication are

   (i) ordinarily light   (ii) laser waves

   (iii) radio-wave   (iv) microwaves

Ans. (ii)

2. Optical fiber communication is based on the phenomenon of

   (i) refraction   (ii) polarization

   (iii) total internal reflection   (iv) diffraction

Ans. (iii)

3. In a single mode fiber, the diameter of core is nearly equal to

   (i) 10 μm   (ii) 50 μm

   (iii) 100 μm   (iv) 125 μm

Ans. (i)

4. The inner most part of an optical fiber is known as

   (i) core   (ii) cladding

   (iii) sheath   (iv) optical fiber axis

Ans. (i)

5. The refractive indices of the materials of core (μ_1) and cladding (μ_2) in an optical fiber satisfy the relation

   (i) $\mu_1 > \mu_2$ (ii) $\mu_1 = \mu_2$

   (iii) $\mu_1 < \mu_2$ (iv) none of them

Ans. (i)

6. In a graded index optical fiber the refractive index of the core

    (i) increases towards the axis of the core.

    (ii) is non-uniform

    (iii) is the same at core-cladding interface

    (iv) decreases towards the axis of the core.

Ans. (i)

7. The acceptance angle in terms of refractive indices of the materials of core (μ_1) and cladding (μ_2), when the end face of an optical fiber is exposed by the air, is

    (i) $\cos^{-1}(\mu_1^2 - \mu_2^2)$

    (ii) $\sin^{-1}(\mu_1^2 - \mu_2^2)$

    (iii) $\cos^{-1}\sqrt{(\mu_1^2 - \mu_2^2)}$

    (iv) $\sin^{-1}\sqrt{(\mu_1^2 - \mu_2^2)}$

Ans. (iv)

8. Suppose d_1 is the diameter of a single mode fiber and d_2 of a multimode fiber. We have

    (i) $d_1 = d_2$ (ii) $d_1 > d_2$ (iii) $d_1 < d_2$ (iv) nothing is definite

Ans. (iii)

9. Suppose α_1 is the acceptance angle for a single mode fiber and α_2 for a multimode fiber. We have

    (i) $\alpha_1 = \alpha_2$ (ii) $\alpha_1 > \alpha_2$ (iii) $\alpha_1 < \alpha_2$ (iv) nothing is definite

Ans. (iii)

12. Problems and questions

1. Describe about various parts of an optical optical fiber.

2. Describe about an optical fiber and explain the terms: (a) Acceptance angle, (b) Acceptance cone, (c) Numerical aperture, (d) Relative refractive index, (e) Propagating modes, and (f) Normalized frequency

3. Write in brief about the advantages of fiber optics as compared to the traditional metal communication lines.

4. What are single mode, multi-mode and graded index fibers? Also, describe in detail the difference in the structures of single mode step index and multi-mode graded index fiber.

5. Discuss the physical significance of numerical aperture. How does it depend on the refractive indices of the core and cladding?

6. Describe the allowed modes of an optical fiber. How they are related to the normalized frequency?

7. Discuss the propagation mechanisms of higher waves in an optical fiber.

8. Describe schematically the basic elements of an optical fiber communication system.

9. Explain why does fraction of power of a signal is lost due to bending of fiber.

10. Describe about the transmission windows for optical fibers.

11. Write notes on the following

   (i) Fiber optics

   (ii) Numerical aperture and its physical significance

   (iii) Core and cladding

   (iv) Acceptance angle

   (v) Single mode step index fiber optics

   (vi) Multi-mode fiber

   (vii) Skip distance

   (viii) Transmission windows for optical fibers

IV. Elasticity

In mechanics, a body is considered as consists of a large number of infinitely small particles closely fixed with respect to one another in three-dimensional space. When relative positions of the particle in a body do not change on application of an external force, the body is said to be rigid body. In true sense, there is no rigid body, as on application of some large amount of force, relative positions of constituent particles would change and the body is said to be deformed. The amount of this applied force would depend on the body. When the applied force is less, the body may be treated as rigid body, for practical purposes.

The amount of deformation in a body depends on the nature of the material of the body and the applied force. The deformation may be classified into two categories: (i) permanent deformation and (ii) temporary deformation. When the deformation remains unchanged after removal of the applied force, the body is termed as plastic and the property of the body by virtue of which it happens so, is called the plasticity. When the retention of deformed shape is complete, the body is said to be perfectly plastic. On the other side, when the deformation is temporary, *i.e.*, the body regains its original shape and size after removal of the external force, the body is termed as elastic and the property of the body by virtue of which it happens so, is called the elasticity. When the body regains its original shape and size instantaneously after removal of the external force, the body is called perfectly elastic. In practice, every body tries to retain its original shape when the applied external force is small and thus shows the elastic property. Also every body lose the property of retaining its shape at least when the applied external force is very large and thus shows the plastic property. Thus, the elastic and plastic properties of a body depend on the magnitude of external force.

The elasticity may be described as the property of a material body by virtue of which the body regains its original shape and size when an external deforming force acting on the body is removed.

1. Stress

The constituent particles in a body are held together by the force of attraction between the particles. When an external force is applied on the body, the force tries to change the relative distances between the particles. This change is opposed by the force of attraction. This force tries to restore the shape and size of the body. Thus, an internal force called the restoring force is developed in the body when an external force is applied on it. In the state of equilibrium, the restoring force is opposite and equal (in magnitude) to the applied force. Now, the restoring force acting per unit area is known as the stress. If F is the external force applied on a body then in the equilibrium state the restoring force developed in the body is F. However, the two forces (the restoring force and the applied force) act in opposite directions to each other. When the stress force is acting on an area A, the stress S is expressed as

$$S = \frac{F}{A}$$

The unit of stress is Newton/meter^2 or N/m^2 in S.I. units. Its dimension is $ML^{-1}T^{-2}$. There are two types of stresses: (i) Normal stress and (ii) tangential stress.

1.1 Normal stress

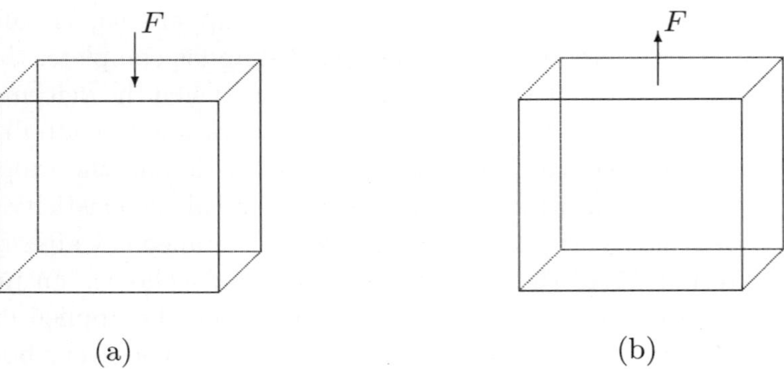

Figure 1: (a) Normal force F working in the inward direction. (b) Normal force F working in the outward direction.

When the applied force (and consequently the restoring force) is working along the normal to the surface area, the stress is known as the normal stress. Figure 1 (a) shows the force acting normal to surface area in the inward direction and Figure 1 (b) shows the force acting normal to surface area in the outward direction. The restoring force is in the direction opposite to the direction of the applied force.

1.2 Tangential stress

When the applied force (and consequently the restoring force) is working along the tangent to the area, the stress is known as the tangential stress. Figure 2 (a) shows the force acting tangential to surface area in the right direction and Figure 2 (b) shows the force acting tangential to surface area in the left direction. The restoring force is in the direction opposite to the direction of the applied force.

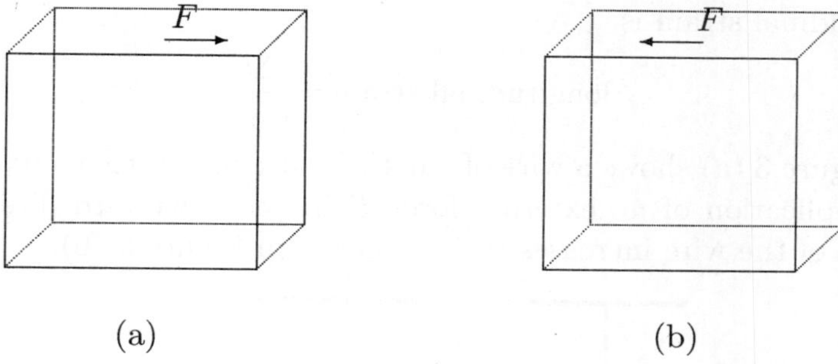

Figure 2: (a) Tangential force F working in the right direction. (b) Tangential force F working in the left direction.

Exercise 1: A force of 45 N is applied on a rectangle surface with length 0.5 m and width 0.3 m of a body. Calculate the amount of stress developed in the body in the equilibrium state.

Solution: Area of the rectangle surface $= 0.5 \times 0.3 = 0.15$ m^2. The strain developed in the body is

$$\text{stress} = \frac{\text{force}}{\text{area}} = \frac{45}{0.15} = 300/\text{N/m}^2$$

It would be normal stress when the applied force is normal to the surface; tangential stress when the applied force is tangential to the surface.

2. Strain

Deformation is produced in a body under the action of an external force. Consequently, the dimensions of the body may change. The strain is defined as change in dimension per unit (original) dimension. Obviously,

the strain is a dimensionless quantity. There are four types of strains:
(i) Longitudinal strain, (ii) Lateral strain, (iii) Volume strain, and (iv)
Shear strain.

2.1 Longitudinal strain

Under the normal stress, the length of a body may be increased or
decreased, depending on the direction of the applied force. (Here, we
shall not account for the change in the lateral direction.) The ratio of
the change in length to the original length is defined as the longitudinal
strain. Suppose, l_0 is original length of a body in absence of an external
force. If on application of an external force (along the length), the length
increases (decreases) to l. The change in length $\Delta l = |l - l_0|$. Thus, the
longitudinal strain is

$$\text{longitudinal strain} = \frac{\Delta l}{l_0}$$

Figure 3 (a) shows a wire of length l_0 suspended from a fixed ceiling.
On application of an external force F in the downward direction, the
length of the wire increases to l, as shown in Figure 3 (b).

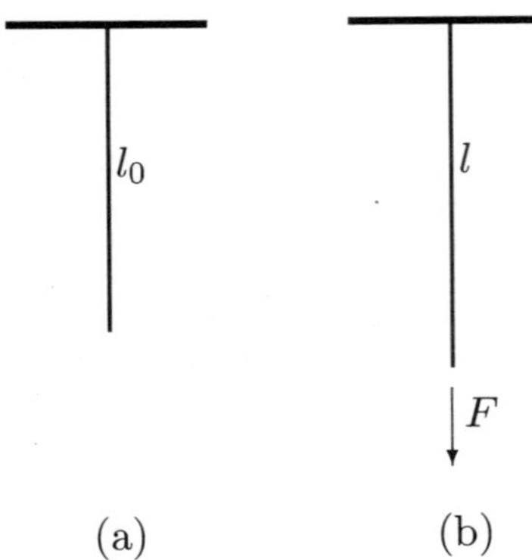

Figure 3: (a) A wire of length l_0 is suspended from a fixed ceiling. (b) On
applying an external force F in the downward direction, the length of the wire
increases to l.

The longitudinal strain per unit normal stress is usually denoted by
the symbol α. Thus, we have

$$\alpha = \frac{\text{longitudinal strain}}{\text{normal stress}}$$

Exercise 2: A wire of length 2.5 m is suspended from a roof. When a
weight of 5 N is applied at the lower end of the wire, the wire is stretched
to the length 2.7 m. Calculate the longitudinal strain.

Solution: Original length $l_0 = 2.5$ m

Change in length $\Delta l = 2.7 - 2.5 = 0.2$ m

$$\text{longitudinal strain} = \frac{\Delta l}{l_0} = \frac{0.2}{2.5} = 0.08$$

Exercise 3: A wire of length $2\,l$ is stretched without tension between the points A and B, as shown in figure. Calculate the longitudinal strain in the wire when it is pulled at the mid-point to the shape ACB.

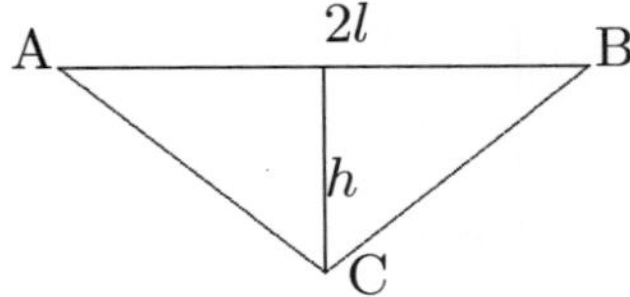

Solution: The original length $l_0 = 2l$. The length AC $= \sqrt{l^2 + h^2}$. Final length of the wire is $2\sqrt{l^2 + h^2}$. The increase in length is

$$\Delta l = 2\sqrt{l^2 + h^2} - 2l = 2l\left[\left(1 + \frac{h^2}{l^2}\right)^{1/2} - 1\right]$$

Using Binomial expansion and neglecting smaller terms, we have

$$\Delta l = 2l\left[1 + \frac{h^2}{2l^2} - 1\right] = \frac{h^2}{l}$$

Therefore, the longitudinal strain is

$$\text{longitudinal strain} = \frac{\Delta l}{l_0} = \frac{h^2/l}{2l} = \frac{h^2}{2l^2}$$

2.2 Lateral strain

It is experimentally true and can be conceived easily that when a body is strained longitudinally, besides the longitudinal change, a lateral change (extension or contraction) takes place. For example, when length of a cylindrical wire is increased by longitudinal stress, its diameter decreases. This change in lateral dimension (perpendicular to the direction of the applied stress) is known as the lateral strain.

Suppose, d_0 is original diameter of a cylindrical body in absence of external force. If on application of an external force along its length, the diameter increases (decreases) to d. The change in diameter $\Delta d = |d - d_0|$. Thus, the lateral strain is

$$\text{lateral strain} = \frac{\Delta d}{d_0}$$

Figure 4 (a) shows a cylinder of length l_0 and diameter d_0 suspended from a fixed ceiling. On application of an external force F in the downward direction, the length of the wire increases to l and diameter decreases to d, as shown in Figure 4 (b).

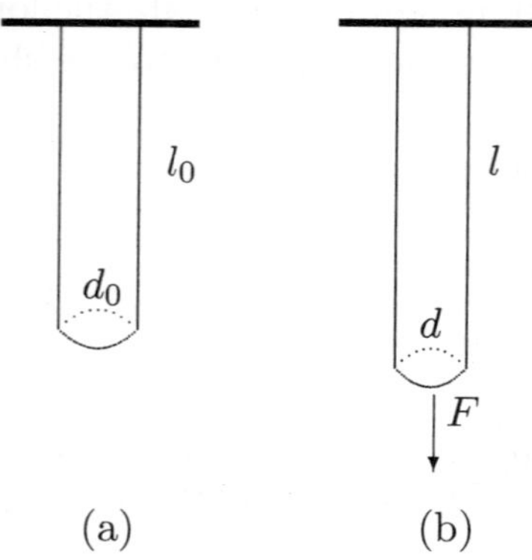

Figure 4: (a) A cylinder of length l_0 and diameter d_0 is suspended from a fixed ceiling. (b) On applying an external force F in the downward direction, the length of the cylinder increases to l and the diameter decreases to d.

The lateral strain per unit normal stress is usually denoted by β. Thus, we have

$$\beta = \frac{\text{lateral strain}}{\text{normal stress}}$$

Note: Notice that the longitudinal strain and lateral strain are of opposite sign to each other. It is so, because when the longitudinal dimension increases, the lateral dimension decreases and vice versa.

Exercise 4: A wire of length 2.5 m and diameter 3 mm is suspended from a roof. When a weight of 5 N is applied at the lower end of the wire, the wire is stretched to the length 2.7 m and its diameter decreases to 2.7 mm. Calculate the lateral strain.

Solution: Original diameter $d_0 = 3$ mm

Change in diameter $\Delta d = |2.7 - 3| = 0.3$ mm

$$\text{lateral strain} = \frac{\Delta d}{d_0} = \frac{0.3}{3} = 0.1$$

2.3 Poisson ratio

For a material, the ratio of lateral strain to the longitudinal strain is a constant quantity and in known as the Poisson ratio, generally denoted by σ. It is dimensionless quantity. Thus, we have

$$\text{Poisson ratio} = \frac{\text{lateral strain}}{\text{longitudinal strain}}$$

Figure 4 (a) shows a cylinder of length l_0 and diameter d_0 suspended from a fixed ceiling. On application of an external force F in the downward direction, the length of the wire increases to l and the diameter decreases to d, as shown in Figure 4 (b). The change in length $\Delta l = l - l_0$ and change in diameter $\Delta d = |d - d_0|$. The longitudinal strain is

$$\text{longitudinal strain} = \frac{\Delta l}{l_0}$$

and lateral strain is

$$\text{lateral strain} = \frac{\Delta d}{d_0}$$

Thus, the Poisson ratio, denoted by σ, is

$$\sigma = \frac{\text{lateral strain}}{\text{linear strain}} = \frac{\Delta d/d_0}{\Delta l/l_0} = \frac{l_0 \Delta d}{d_0 \Delta l}$$

It is a dimensionless quantity.

Exercise 5: A cylindrical rod of length 2.5 m and diameter 3 mm is suspended from a roof. When a weight of 5 N is applied at the lower end of the rod, the rod is stretched to the length 2.7 m and its diameter decreases to 2.7 mm. Calculate the Poisson ratio.

Solution: Original length $l_0 = 2.5$ m

Change in length $\Delta l = 2.7 - 2.5 = 0.2$ m

$$\text{longitudinal strain} = \frac{\Delta l}{l_0} = \frac{0.2}{2.5} = 0.08$$

Original diameter $d_0 = 3$ mm

Change in diameter $\Delta d = |2.7 - 3| = 0.3$ mm

$$\text{lateral strain} = \frac{\Delta d}{d_0} = \frac{0.3}{3} = 0.1$$

Thus, we have

$$\text{Poisson ratio} = \frac{\text{lateral strain}}{\text{longitudinal strain}} = \frac{0.1}{0.08} = 1.25$$

Exercise 6: A cylindrical wire of length L and diameter D is stretched by applying an external force. If the increase in length of the wire is l and the Poisson ratio of the material of the wire is σ, find the change in the volume of the wire.

Solution: The original volume of the wire is

$$V = \frac{\pi D^2}{4}\, L$$

On application of an external force, suppose d is the change in diameter. We have the Poisson ratio σ as

$$\sigma = \frac{l/L}{d/D} \qquad \text{or} \qquad d = \frac{lD}{L\sigma}$$

The final length of the wire $L' = L + l$ and diameter D' is

$$D' = D - d = D - \frac{lD}{L\sigma} = D\left[1 - \frac{l}{L\sigma}\right]$$

The final volume V' of the wire

$$V' = \frac{\pi D'^2}{4}\, L' = \frac{\pi(L+l)D^2}{4}\left[1 - \frac{l}{L\sigma}\right]^2$$

$$= \frac{\pi D^2}{4}\left[L - \frac{2l}{\sigma} + l\right]$$

Thus, the change in volume is

$$\Delta V = \frac{\pi D^2}{4}\left[L - \frac{2l}{\sigma} + l\right] - \frac{\pi D^2}{4}\, L = \frac{\pi D^2 l}{4}\left[1 - \frac{2}{\sigma}\right]$$

Exercise 7: A material has Poisson ratio $\sigma = 0.2$. If a uniform rod made of this material suffers longitudinal strain 4×10^{-3}, calculate the percentage change in its volume.

Solution: longitudinal strain $= 4 \times 10^{-3}$ and Poisson ratio $\sigma = 0.2$. We have

$$\text{lateral strain} = \sigma(\text{longitudinal strain}) = 0.2(4 \times 10^{-3}) = 8 \times 10^{-4}$$

Suppose V be the initial volume of a rod of length l and radius r, then

$$V = \pi r^2 l$$

On differentiation, we get the change in volume

$$\Delta V = \pi r^2 \Delta l + 2\pi r l \Delta r$$

Thus, we have

$$\frac{\Delta V}{V} = \frac{\pi r^2 \Delta l}{\pi r^2 l} + \frac{2\pi r l \Delta r}{\pi r^2 l} = \frac{\Delta l}{l} + \frac{2\Delta r}{r}$$

Suppose the length of the rod increases. We have longitudinal strain

$$\frac{\Delta l}{l} = 4 \times 10^{-3}$$

and lateral strain is

$$\frac{\Delta r}{r} = -8 \times 10^{-4}$$

The minus sign is used as the longitudinal strain and lateral strain are of opposite sign to each other. Using these values, we have

$$\frac{\Delta V}{V} = 4 \times 10^{-3} + 2(-8 \times 10^{-4}) = 2.4 \times 10^{-3}$$

The percentage change in its volume is

$$\frac{\Delta V}{V} \times 100 = (2.4 \times 10^{-3})100 = 0.24$$

2.4 Volume strain

Suppose equal stresses are applied on a body form all directions. The volume of the body increases or decreases, depending on the directions of stresses (*i.e.*, inward or outward direction) without change in its shape. The ratio of change in volume of the body to its original volume is known as the volume strain. Suppose, V_0 is original volume of a body in absence of external forces. If on application of external forces, the volume changes to V. The change in volume $\Delta V = |V - V_0|$. Thus, the volume strain is

$$\text{volume strain} = \frac{\Delta V}{V_0}$$

Figure 5 (a) shows a body having volume V_0 placed in the space. On application of inward external forces from all directions, the volume of the body decreases to V, as shown in Figure 5 (b).

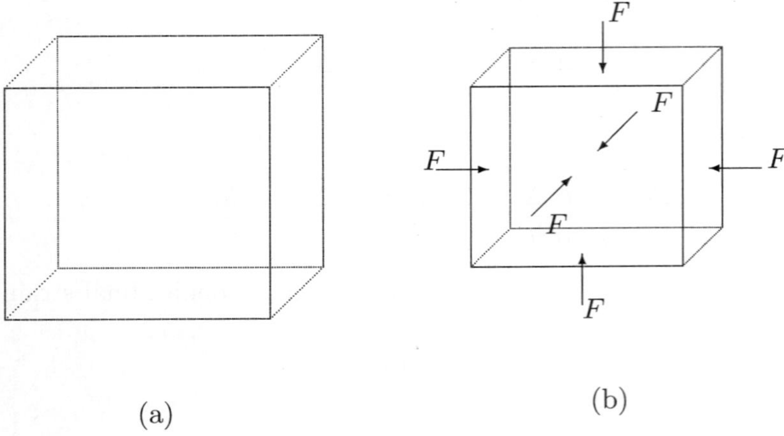

Figure 5: (a) A Body of volume V_0. (b) On applying equal inward external forces from all directions, the volume of the body decreases to V.

Exercise 8: A parallelepiped has length 0.6 m, width 0.4 m and height 0.3 m. When external forces are applied on the parallelepiped from all directions, its length, width and height, receptively reduce to 0.55 m, 0.35 m, 0.25 m. Calculate the volume strain.

Solution: For a parallelepiped, we have

Original volume $V_0 = 0.6 \times 0.4 \times 0.3 = 0.072$ m^3.

Final volume $V = 0.55 \times 0.35 \times 0.25 = 0.048125$ m^3.

Change in volume $\Delta V = |0.048125 - 0.072| = 0.023875$ m^3.

Thus, the volume strain is

$$\text{volume strain} = \frac{\Delta V}{V_0} = \frac{0.023875}{0.072} = 0.3183$$

2.5 Shear strain

In order to understand shear strain, let us imagine a cubic volume whose one face is fixed on a horizontal platform and its opposite face (the face at the top) is subjected to a tangential force as shown in Figure. It can easily be conceived that in equilibrium position, the shape of the cube becomes as shown by dotted lines. The tangential force applied on the upper plane produces a reaction force on the fixed plane (plane at the bottom) which is tangential to the plane. Under these two forces, a shear stress is produced through out the material of the body which makes the different layers of the cube to slip relative to its lower layer in the direction of the applied force. Thus, the shape of the cube is deformed, but its volume does not change.

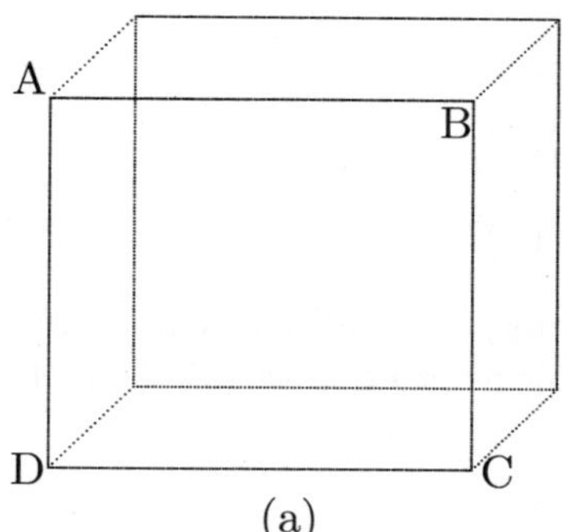
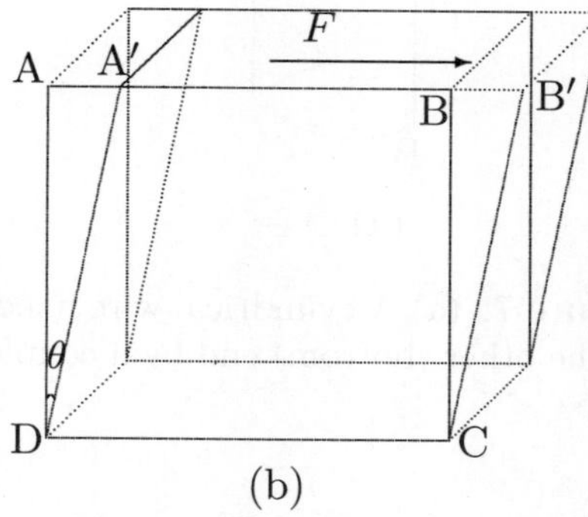

Figure 6: (a) A cube in absence of any external force. (b) Deformation of a cube under shear strain.

Figure 6 (a) shows a cube placed on some horizontal surface. On application of an external force F tangential to the top surface and keeping the bottom surface fix, the cube is deformed. For example, the front surface ABCD is deformed to A'B'CD, as shown in Figure 6 (b).

The shift in the relative position of the plane at a unit distance apart in the direction of the force when the body is in the equilibrium position is termed as the shear strain.

$$\text{shear strain} = \frac{\text{AA}'}{\text{AD}} = \tan\theta$$

Sometimes θ is referred to as shear angle instead of shear strain.

In order to make meaning of shear angle more clear, let us consider another example. Suppose, we have a cylindrical wire fixed at one (top) end, as shown in Figure 7 (a). The wire is twisted at the other (bottom) end by a couple acting in the cross sectional plane of the wire. The wire will be twisted and each point on its lower cross section will be shifted from its original position. For example, the point B in the circumference of its lower cross sectional plane will move to B'. Then we have

$$\text{shear strain} = \frac{\text{BB}'}{\text{AB}} = \theta$$

when the angle θ is small (BB$'$ $<<$ AB).

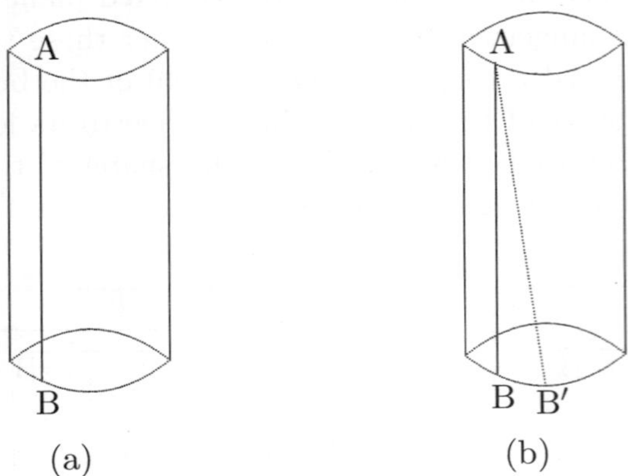

Figure 7: (a) A cylindrical wire fixed at one (top) end. (b) The wire is twisted at the other (bottom) end by a couple acting in the cross sectional plane of the wire.

3. Variation of strain with stress and elastic limit

When stretching stress is applied on a wire, its length increases. The variation of strain with stress for most of the materials is as shown in Figure. Initially, when the stress is increased, the strain varies linearly as shown by OA in the figure. The point A determines the limit of proportionality. Up to this point, the strain is proportional to the stress in equilibrium position (Hooke's law).

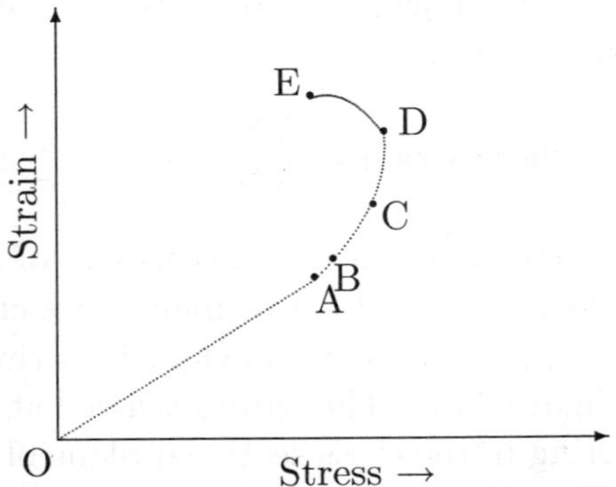

Figure 8: Variation of strain versus stress and elastic limit.

Close to the point A, there is a point B which is termed as the point of elastic limit. In many case the points A and B coincide, but in some cases they are separate points. In the narrow range AB, the strain increases with a little more rapid rate case compared to that in the straight part OA. When the stress lies in the range OB, complete recovery of the original shape and size of the body is obtained when the applied force is removed. When the stress is increased beyond the elastic limit (point B), the strain-stress curve bends towards the strain-axis. It shows that the rate of increase of strain becomes much faster with the stress until the point C is reached. This point C is known as the yield point. Once the stress is reached up to this point, the wire appear to flow and it length increases without increasing the applied force. Beyond this point, the wire suffers plastic deformation. This part CD of the curve is almost parallel to strain-axis. Once the point D is reached, even little unloading of the wire is not able to stop the flow of wire. The thinning of wire which apparently starts appearing at the point C eventually develops and the point E is reached. Soon after the wire breaks. The point E is known as breaking point. The tensile stress at this point determines the ultimate strength of the material of the wire.

4. Hooke's law and coefficient of elasticity

It was concluded in 1679 by Hooke that within the elastic limit, strain is directly proportional to the stress. This is known as the Hooke's law. Thus, we have

$$\text{stress} \propto \text{strain} \qquad \text{or} \qquad \frac{\text{stress}}{\text{strain}} = E \text{ (constant)}$$

This constant E is called the modulus of elasticity. As the strain does not have any unit, hence the E has the same unit as that of the stress. The modulus of elasticity is of three types, depending on the type of strain produced. These are: (i) Young's modulus, (ii) Bulk modulus and (iii) Modulus of rigidity.

4.1 Young's modulus

The Young's modulus is defined as the ratio of normal stress to the linear strain. Suppose a wire of length l and cross sectional area A is subjected to a force F in the direction of the normal to the area of cross section. It is denoted by Y and its unit is Newton per square meter (N/m^2). Let the length increases by Δl. The normal stress is

$$\text{normal stress} = \frac{F}{A}$$

and the linear strain is

$$\text{linear strain} = \frac{\Delta l}{l_0}$$

Hence, the Young's modulus is

$$Y = \frac{\text{normal stress}}{\text{linear strain}} = \frac{F/A}{\Delta l / l_0} = \frac{F l_0}{A \Delta l}$$

Exercise 9: A wire having length of 2.5 m and area of cross-section 2.3 mm^2 is stretched by a weight of 2 kg. Calculate the increase in length of the wire. Young's modulus of the material of the wire is 1.8×10^{11} N/m^2.

Solution: We have $l_0 = 2.5$ m, $A = 2.3$ mm$^2 = 2.3 \times 10^{-6}$ m^2, $F = mg = 2 \times 9.8 = 19.6$ N, $Y = 1.8 \times 10^{11}$ N/m^2. For the increase Δl in length, we have the relation

$$Y = \frac{F l_0}{A \Delta l} \qquad \text{therefore} \qquad \Delta l = \frac{F l_0}{AY}$$

On using the values, we get

$$\Delta l = \frac{19.6 \times 2.5}{(2.3 \times 10^{-6})(1.8 \times 10^{11})} = 1.18 \times 10^{-4} \text{ m} = 0.118 \text{ mm}$$

4.2   Bulk modulus

When a body is subjected to equal forces from all directions, perpendicular to its surface area, then due to the compressing forces, the volume of the body decreases without changing its shape. Similarly, when equal stretching forces are applied in from all directions, perpendicular to its surfaces, then the volume of the body increases without changing its shape. The ratio of this type of forces acting per unit surface area to the volume strain is known as the Bulk modulus. It is usually denoted by K and its unit is Newton per square meter (N/m^2). The force per unit area is known as the pressure, denoted by P. For the original volume V_0, suppose ΔV denotes the change in the volume. The volume strain is expressed as

$$\text{volume strain} = \frac{\Delta V}{V_0}$$

Hence, the Bulk modulus K is

$$K = \frac{\text{pressure}}{\text{volume strain}} = \frac{P}{\Delta V / V_0} = \frac{P V_0}{\Delta V}$$

4.2.1 Compressibility

The reciprocal of the Bulk modulus is known as the compressibility. It is usually denoted by B. Therefore,

$$B = \frac{1}{K} = \frac{\Delta V}{PV_0}$$

Exercise 10: A solid sphere of radius 12 cm is subjected to a uniform pressure equal to 4×10^8 N/m^2. Calculate the change in volume of the sphere. The Bulk modulus of the material of the sphere is 3.25×10^{11} N/m^2.

Solution: We have pressure $P = 4 \times 10^8$ N/m^2, radius $r = 12$ cm $= 12 \times 10^{-2}$ m, Bulk modulus $K = 3.25 \times 10^{11}$ N/m^2. volume

$$V_0 = \frac{4}{3}\pi r^3 = \frac{4}{3}(3.14)(12 \times 10^{-2})^3 = 7.23 \times 10^{-3} \text{ m}^3$$

For the increase ΔV in volume, we have

$$K = \frac{PV_0}{\Delta V} \qquad \text{or} \qquad \Delta V = \frac{PV_0}{K}$$

On using the values, we have

$$\Delta V = \frac{(4 \times 10^8)(7.23 \times 10^{-3})}{3.25 \times 10^{11}} = 8.90 \times 10^{-6} \text{ m}^3$$

Exercise 11: The compressibility of water is 4×10^{-5} per unit atmospheric pressure. Calculate the decrease in volume of 1 liter of water under the pressure of 100 atmosphere. Given, 1 atmospheric pressure $= 1.013 \times 10^5$ N/m^2.

Solution: Pressure $P = 100$ atm $= 100\,(\,1.013 \times 10^5\,) = 1.013 \times 10^7$ N/m^2. The Bulk modulus K is

$$K = \frac{1}{B} = \frac{1}{4 \times 10^{-5}} = 2.5 \times 10^4 \text{ atm} = (2.5 \times 10^4)(1.013 \times 10^5)$$

$$= 2.5325 \times 10^9 \text{ N/m}^2$$

Initial volume $V_0 = 1$ liter $= 1000$ cm$^3 = 10^{-3}$ m^3
The compressibility B is

$$K = \frac{PV_0}{\Delta V} \qquad \text{or} \qquad \Delta V = \frac{PV_0}{K}$$

where ΔV is the change in volume. Using the values, we have

$$\Delta V = \frac{(1.013 \times 10^7)(10^{-3})}{2.5325 \times 10^9} = 4 \times 10^{-6} \text{ m}^3$$

4.3 Modulus of rigidity

Under the shearing force, the shape of a body can be deformed without changing its volume. The shear strain is the measure of this deformation. The ratio of shear stress to shearing strain is known as the modulus of rigidity and is denoted by η. The shear strain is measured in terms of the angle. Thus, if S be the shearing stress and θ the shear strain, we have

$$\eta = \frac{S}{\theta}$$

The unit of η is that of stress.

5. Equivalence to shear strain to compressive and extension strains

In order to prove this equivalence, let us consider a cube whose each arm is of length L. Suppose, the cube is subjected to a tangential force F along its face at the top in the direction DC, as shown in Figure 9.

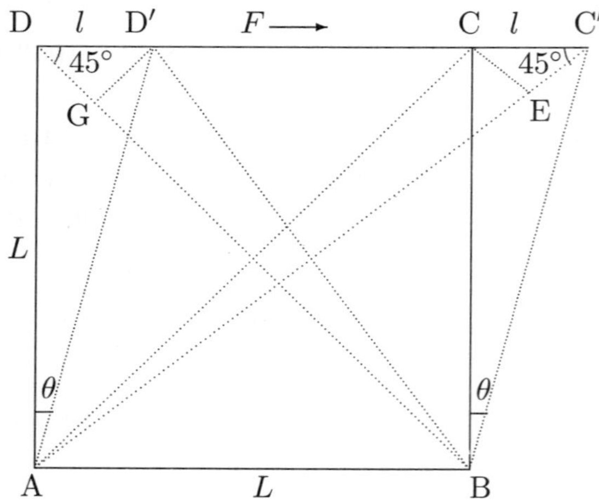

Figure 9: Force F is applied along DC.

The shear strain is

$$\theta = \frac{DD'}{AD} = \frac{CC'}{BC}$$

Here, we have used the fact that the angle θ is small. On application of the tangential for F, we find that the diagonal AC is extended to AC' and the diagonal BD is contracted to BD'. Draw perpendicular CE on AC'. Thus, the extension in AC (= AE) is EC'. Draw perpendicular GD' on BD. Thus, the contraction in BD is DG. Now, the extension strain along AC can be written as

$$\frac{\text{increase in length}}{\text{original length}} = \frac{\text{EC}'}{\text{AE}} = \frac{\text{EC}'}{\text{AC}}$$

As the angle θ is very small, the angle CC'B is nearly $90°$ and the angle CC'E is nearly $45°$. Thus, from the triangle CC'E, we have

$$\text{EC}' = \text{CC}' \cos 45° = \frac{\text{CC}'}{\sqrt{2}}$$

In triangle ABC, we have

$$\frac{\text{BC}}{\text{AC}} = \sin 45° = \frac{1}{\sqrt{2}} \qquad \text{or} \qquad \text{AC} = \text{BC}\sqrt{2} = L\sqrt{2}$$

Thus, we have

$$\frac{\text{EC}'}{\text{AC}} = \frac{\text{CC}'}{\sqrt{2}} / \text{BC}\sqrt{2} = \frac{1}{2} \frac{\text{CC}'}{\text{BC}} = \frac{\theta}{2}$$

Similarly, the compression strain along BD can be written as

$$\frac{\text{decrease in length}}{\text{original length}} = \frac{\text{DF}}{\text{BD}}$$

As the angle θ is very small, the angle DD'A is nearly $90°$ and the angle D'DF is nearly $45°$. Thus, from the triangle DD'F, we have

$$\text{DF} = \text{DD}' \cos 45° = \frac{\text{DD}'}{\sqrt{2}}$$

In triangle ABD, we have

$$\frac{\text{AD}}{\text{BD}} = \sin 45° = \frac{1}{\sqrt{2}} \qquad \text{or} \qquad \text{BD} = \text{AD}\sqrt{2} = L\sqrt{2}$$

Thus, we have

$$\frac{\text{DF}}{\text{BD}} = \frac{\text{DD}'}{\sqrt{2}} / \text{AD}\sqrt{2} = \frac{1}{2} \frac{\text{DD}'}{\text{AD}} = \frac{\theta}{2}$$

Thus, we have

shear strain = extension strain + compression strain

This shows that total shear strain is a result of two tensile strains perpendicular to each other; one corresponding to extension and the other corresponding to compression.

6. Relations between elastic constants

The three kinds of elastic coefficients, namely, Young's modulus Y, Bulk modulus K, modulus of rigidity η, together with the Poisson ratio σ are known as the elastic constants. These constants are related to one another through some relations. In order to get the relations among them, let us consider a cube of length L consists of isotropic medium. The volume of the cube is L^3. Suppose, F_x, F_y, F_z are the forces acting perpendicular to its faces and along the x, y, z axes, respectively. Let us assume that the magnitudes of all these forces are equal

$$F_x = F_y = F_z = F \text{ (say)}$$

and therefore, we have the stresses

$$S_x = S_y = S_z = S \text{ (say)}$$

The directions of S_x, S_y, S_z are along Ox, Oy, Oz, respectively, as shown in Figure 10. Each of these forces will produce extension along its direction and contraction in other two directions, perpendicular to it. The extension produced by the force F_x along x-direction is SL/Y and and each of the contractions produced by it along y and z directions is $\sigma SL/Y$.

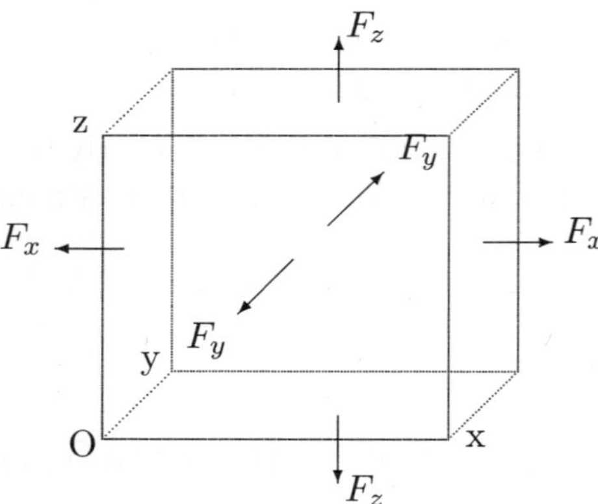

Figure 10: Equal forces are acting in all directions in the outward direction.

Similarly, the extension produced by the force F_y along y-direction is SL/Y and each of the contractions produced by it along x and z directions is $\sigma SL/Y$. Further, the extension produced by the force F_z along z-direction is SL/Y and each of the contractions produced by it along x and y directions is $\sigma SL/Y$. The net extension produced, by all the three forces, along x-direction is

$$\frac{SL}{Y} - \frac{\sigma SL}{Y} - \frac{\sigma SL}{Y} = \frac{SL}{Y}(1 - 2\sigma)$$

Similarly, the net extension produced by all the three forces along y-direction is

$$\frac{SL}{Y}(1 - 2\sigma)$$

Also, the net extension produced by all the three forces along z-direction is

$$\frac{SL}{Y}(1 - 2\sigma)$$

Therefore, under the action of all the three forces, the length of each arm of the cube becomes

$$L + \frac{SL}{Y}(1 - 2\sigma) = L\left[1 + \frac{S}{Y}(1 - 2\sigma)\right]$$

Thus, the volume of the cube is

$$\text{volume} = L^3\left[1 + \frac{S}{Y}(1 - 2\sigma)\right]^3 = L^3\left[1 + \frac{3S}{Y}(1 - 2\sigma)\right]$$

Here, in the expansion, the higher order terms have been neglected, as they are very small. Hence, the change in volume is

$$\text{change in volume} = L^3\left[1 + \frac{3S}{Y}(1 - 2\sigma)\right] - L^3 = \frac{3SL^3}{Y}(1 - 2\sigma)$$

The volume strain is

$$\text{volume strain} = \frac{\text{change in volume}}{\text{original volume}} = \frac{(3SL^3/Y)(1 - 2\sigma)}{L^3} = \frac{3S}{Y}(1 - 2\sigma)$$

The Bulk modulus K of the material of the cube is

$$K = \frac{\text{stress}}{\text{volume strain}} = \frac{S}{(3S/Y)(1 - 2\sigma)} = \frac{Y}{3(1 - 2\sigma)}$$

Thus, we have

$$Y = 3K(1 - 2\sigma) \tag{4.1}$$

This is relation between Y, K and σ.

In order to derive relation between Y, η and σ, let us consider a cube where tangential force F is applied on the top surface ABHG, as shown in Figure 11. As the consequence, this face shifts to the new position A′B′H′G′. The shear strain produced is

$$\theta = \frac{\text{AA}'}{L} = \frac{\text{BB}'}{L}$$

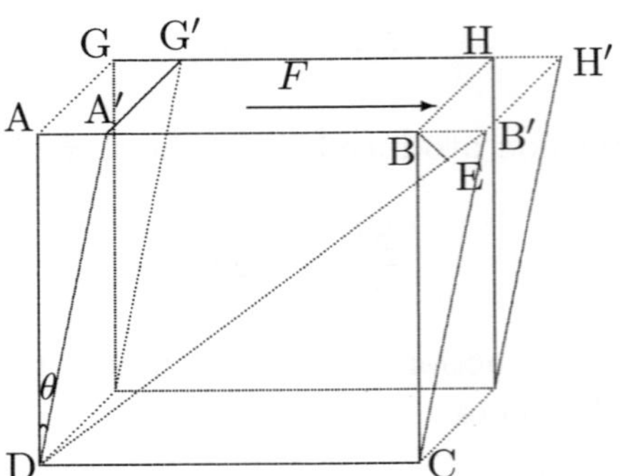

Figure 11: Cube subjected to a tangential force F applied on the top surface ABHG. The bottom surface is kept fix.

Let us consider the face ABCD. The force F is acting along the arm AB. This force will produce a stress $S = F/L^2$ along AB. This stress is equivalent to linear tensile stress of same magnitude along the diagonal BD and equal to compressive stress along the diagonal AC which are perpendicular to each other. Both these stresses will produce extension in the diagonal DB.

$$\text{extension produced due to tensile stress} = \frac{\text{DB } S}{Y} = \frac{\sqrt{2}LS}{Y}$$

$$\text{extension produced due to compressive stress} = \frac{\text{DB}\sigma S}{Y} = \frac{\sqrt{2}L\sigma S}{Y}$$

$$\text{Total extension produced in the diagonal DB} = \frac{\sqrt{2}LS}{Y} + \frac{\sqrt{2}L\sigma S}{Y}$$

$$= \frac{\sqrt{2}LS(1+\sigma)}{Y}$$

Let us evaluate BB'. For this draw perpendicular BE on DB' as shown in Figure 11. We have

$$\text{BB}' = \frac{\text{EB}'}{\cos \text{BB}'\text{E}}$$

Practically, BB' is very small in comparison to AB and in such case, we may assume that $\angle$ BB'E = 45°. Further, B'E is the extension in the length of the diagonal DB. Thus, we have

$$\text{BB}' = \frac{\sqrt{2}LS(1+\sigma)/Y}{\cos 45°} = \frac{2LS(1+\sigma)}{Y}$$

The shear strain θ is

$$\theta = \frac{BB'}{BC} = \frac{2LS(1+\sigma)/Y}{L} = \frac{2S(1+\sigma)}{Y}$$

The modulus of rigidity η is

$$\eta = \frac{\text{shear stress}}{\text{shear strain}} = \frac{S}{2S(1+\sigma)/Y} = \frac{Y}{2(1+\sigma)}$$

Thus, we have

$$Y = 2\eta(1+\sigma) \qquad\qquad (4.2)$$

This is relation between Y, η and σ. These relations (4.1) and (4.2) can be used for getting other relations.

On elimination of Y from equations (4.1) and (4.2), we have

$$3K(1-2\sigma) = 2\eta(1+\sigma) \qquad \text{or} \qquad \sigma = \frac{3K-2\eta}{6K+2\eta} \qquad (4.3)$$

On elimination of σ from equations (4.1) and (4.2), we have

$$\frac{Y}{2\eta} - 1 = \frac{1}{2} - \frac{Y}{6K} \qquad \text{or} \qquad Y = \frac{9K\eta}{3K+\eta} \qquad (4.4)$$

Thus, we get four relations (4.1), (4.2), (4.3), (4.4) between four elastic constants.

Exercise 12: Show that the value of the Poisson ratio σ may lie between -1 and $1/2$.

Solution: We have the relations

$$Y = 3K(1-2\sigma) \qquad \text{and} \qquad Y = 2\eta(1+\sigma)$$

where the Young's modulus Y, Bulk modulus K, modulus of rigidity η all are positive quantities. From these relations, we have

$$3K(1-2\sigma) = 2\eta(1+\sigma)$$

Suppose the Poisson ratio σ is positive, we must have

$$(1-2\sigma) > 0 \qquad \text{or} \qquad \sigma < 1/2$$

Suppose the Poisson ratio σ is negative, we must have

$$(1+\sigma) > 0 \qquad \text{or} \qquad \sigma > -1$$

These two relations for the values of σ say that σ must lie between -1 and $1/2$.

Exercise 13: The Young's modulus Y and modulus of rigidity η of material of a rod are 22×10^{10} N m^{-2} and 9×10^{10} N m^{-2}, respectively. Find the Poisson ratio σ and Bulk modulus K for the material of the rod.

Solution: We know

$$Y = 2\eta(1 + \sigma) \qquad \text{or} \qquad \sigma = \frac{Y}{2\eta} - 1$$

Using the values, we have

$$\sigma = \frac{22 \times 10^{10}}{2 \times 9 \times 10^{10}} - 1 = 0.22$$

We have

$$K = \frac{Y}{3(1 - 2\sigma)} = \frac{22 \times 10^{10}}{3(1 - 2 \times 0.22)} = 13.10 \times 10^{10} \text{ N m}^{-2}$$

Exercise 14: The length of a wire of radius 1 mm increased from 100 cm to 100.5 cm when stretched by a stress of 5×10^8 N m^{-2}. The Poisson ratio σ for the material of the wire is 0.25. Find Young's modulus Y and modulus of rigidity η for the material of the wire. Also find out decrease in the radius of the wire when stretched.

Solution: The linear strain produced is

$$\text{linear strain} = \frac{\Delta l}{l_0} = \frac{100.5 - 100}{100} = 0.005$$

The Young's modulus Y is

$$Y = \frac{\text{normal stress}}{\text{linear strain}} = \frac{5 \times 10^8}{0.005} = 10^{11} \text{ N m}^{-2}$$

The modulus of rigidity η is

$$\eta = \frac{Y}{2(1 - \sigma)} = \frac{10^{11}}{2(1 - 0.25)} = 6.67 \times 10^{10} \text{ N m}^{-2}$$

Suppose the change in radius is Δr. Then

$$\text{lateral strain} = \frac{\Delta r}{r_0} = \frac{\Delta r}{1}$$

We know, the Poisson ratio σ is

$$\sigma = \frac{\text{lateral strain}}{\text{linear strain}} = \frac{\Delta r}{0.005} = 200\Delta r$$

Using the value of σ, we have

$$0.25 = 200\Delta r \qquad \text{and therefore} \qquad \Delta r = 1.25 \times 10^{-3} \text{ mm}$$

7. Bending moment of a beam

A rod or bar consists of isotropic elastic material with uniform area of cross-section whose length is very large in comparison to its diameter or thickness is known as a beam. A beam can be considered as consists of a large number of thin filaments of very small thickness and of same length as that of the beam. Such a beam can be bent by applying forces across its ends. In a bent position, the upper surface of the beam becomes convex and the lower surface becomes concave, as shown in Figure 12.

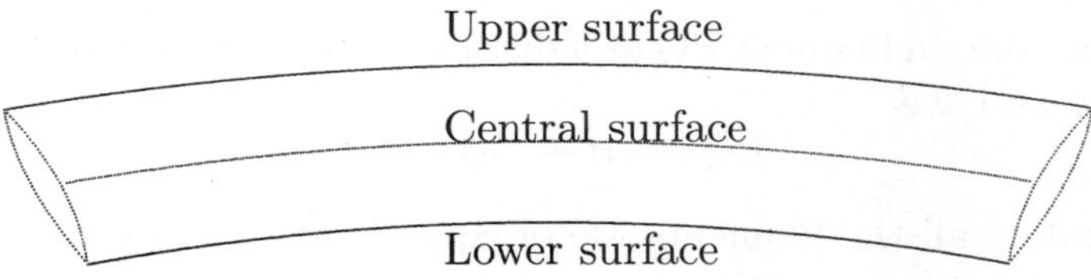

Figure 12: A beam in the bent situation. The upper surface of the beam becomes convex and the lower surface becomes concave.

It is obvious that in the bending position, the filaments lying above the central surface are extended and those lying below the central surface are contracted. This extension and contraction increases with the distance from the central surface. The extension becomes the maximum at the upper surface and the contraction becomes the maximum at the lower surface. The central surface is neither extended nor contracted, and is known as the neutral surface.

Let us consider an element of the beam, we get a view as shown in Figure 13. In this figure, CD is the neutral filament whose length is not changed. Above this, the filaments are extended and below this, the filaments are contracted. Since there is neither shearing stress nor any change in volume of the material, the extensions and contractions of filaments should be due to the forces acting along the length of the filament. Thus, the outward extension forces act on the filament lying above CD and the inner contraction forces on the filaments lying below

CD. Let the filament CD has acquired shape of an arc of a circle of radius R and center at O. Then, CD $= R\phi$.

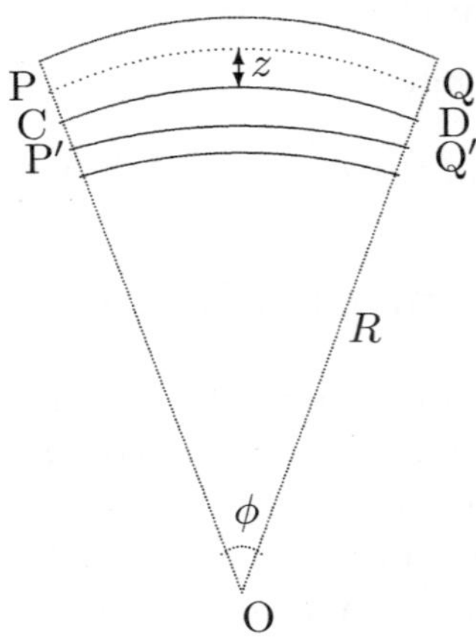

Figure 13: An element of a bent beam. The line CD belongs to the neutral surface.

Now, consider a filament PQ at a distance z from the neutral filament CD. Then, we have

$$\text{PQ} = (R + z)\phi$$

Before bending all the filaments are of equal length. Thus, the increase in length of the filament PQ is

$$\Delta l = \text{PQ} - \text{CD} = (R + z)\phi - R\phi = z\phi$$

Hence, the linear strain produced in the filament is

$$\text{linear strain} = \frac{\Delta l}{\text{CD}} = \frac{z\phi}{R\phi} = \frac{z}{R}$$

Similarly, we can find that the linear strain produced in a filament P$'$Q$'$ situated at a distance z on the other side of the neutral surface is z/R. It shows that the linear strain produced in a filament is proportional to its distance from the neutral surface. The tensile (or contraction) stress required to produce linear strain z/R is

$$\text{normal stress} = \text{Young's modulus} \times \text{linear strain} = Y\,\frac{z}{R} = \frac{Yz}{R}$$

Now consider an area Δa (whose plane if perpendicular to the plane of the paper) at a distance z from the neutral surface. Thus, the force acting on this area is

$$\text{force} = \text{normal stress} \times \text{area} = \frac{Yz}{R}\,\Delta a = \frac{Yz\Delta a}{R}$$

This force will act on the filament in the outward direction (in the plane of the paper) when the filament is above the neutral surface and in the inward direction (in the plane of the paper) when the filament is below the neutral surface. The moment of this force about the neutral filament is

$$\text{moment of force} = \text{force} \times \text{perpendicular distance}$$

$$= \frac{Yz\Delta a}{R}\, z = \frac{Yz^2\Delta a}{R}$$

The moments of all such forces about the neutral filament will be in he same direction. Thus, the total moment of forces acting on all the filaments of the beam is

$$\text{total moment of forces} = \sum \frac{Yz^2\Delta a}{R} = \frac{Y}{R}\sum \Delta a z^2$$

This is known as the bending moment of the beam. There will be a restoring couple which brings equilibrium in the beam. The quantity $\sum \Delta a z^2$ is known as the geometrical moment of inertia of the beam and is denoted by I_g. We have

$$\text{bending moment of the beam} = \frac{YI_g}{R}$$

The quantity YI_g is known as the flexural rigidity of the beam.

(i) For a beam having circular cross section of radius r, we have

$$I_g = \frac{\pi r^4}{4}$$

(ii) For a beam having rectangular cross section of breadth b and thickness d, we have

$$I_g = \frac{bd^3}{12}$$

Exercise 15: A circular hollow rod and a circular solid rod made of the same material have the same length and mass. Which one of them will be more difficult to bend.

Solution: Suppose M and L are mass and length of two rods. Let r be the radius of solid rod, and r_0 and r_1 be inner and outer radii of hollow rod. If ρ be the density of the material of the rod, we have

$$M = \pi r^2 L\rho \qquad \text{and} \qquad M = \pi(r_0^2 - r_1^2)L\rho$$

From these relations, we get

$$\pi(r_0^2 - r_1^2)L\rho = \pi r^2 L\rho \qquad \text{or} \qquad r_0^2 - r_1^2 = r^2 \qquad (4.5)$$

The bending of beam depends on the flexural rigidities of rods which is equal to YaK^2, where Y is the Young's modulus of the material of the rods, a the area of cross-section and K the radius of gyration about its axis.

For the solid rod, we have geometrical moment of inertia

$$I_g = \frac{\pi r^4}{4}$$

Therefore, the flexural rigidity F_s for solid rod is

$$F_s = \frac{Y\pi r^4}{4}$$

For the hollow rod, we have geometrical moment of inertia

$$I_g = \frac{(r_0^4 - r_1^4)}{2}$$

Therefore, the flexural rigidity F_h for hollow rod is

$$F_h = \frac{Y\pi(r_0^4 - r_1^4)}{2}$$

From equation (6.12), we have

$$r_0^2 - r_1^2 = r^2 \qquad \text{and therefore} \qquad r_0^2 + r_1^2 > r^2$$

Hence, we have

$$(r_0^2 - r_1^2)(r_0^2 + r_1^2) > r^4$$

Consequently, we have

$$\frac{r_0^4 - r_1^4}{2} > \frac{r^4}{4} \qquad \text{and therefore} \qquad \frac{Y\pi(r_0^4 - r_1^4)}{2} > \frac{Y\pi r^4}{4}$$

It shows that the flexural rigidity of the hollow rod is larger than that the solid rod. Hence, it would be more difficult to bend the hollow rod as compared to the solid rod.

Exercise 16: Show that a beam of square cross section is more stiff as compared to that with circular cross section when both of them have the same material, mass and length.

Solution: Let us consider two rods A and B of square cross section and circular cross section, respectively of the same material of density ρ. Let the mass and length of the rods are M and L, respectively. Suppose a is the area of cross section of A and r the radius of cross section of B. We have

$$M = aL\rho \qquad \text{and} \qquad M = \pi r^2 L\rho$$

Thus, we have

$$aL\rho = \pi r^2 L\rho \qquad \text{and} \qquad a = \pi r^2$$

The stiffness of a rod is decided with the help of its flexural rigidity YaK^2, where Y is the Young's modulus of the material of the rods, a the area of cross-section and K the radius of gyration about its axis.

For a rod with square cross section, the area of cross section is a and $K^2 = a/12$. Thus, the flexural rigidity F_s of square rod is

$$F_s = Ya\,\frac{a}{12} = \frac{Ya^2}{12}$$

For a rod with circular cross section, the area of cross section is $a = \pi r^2$ and $K^2 = r^2/4$. Thus, the flexural rigidity F_c of circular rod is

$$F_c = Y\pi r^2\,\frac{r^2}{4} = \frac{Y\pi r^4}{4}$$

Using relation $a = \pi r^2$, we have

$$F_s = \frac{Y(\pi r^2)^2}{12} = \frac{Y\pi^2 r^4}{12}$$

We have

$$\frac{F_s}{F_c} = \frac{Y\pi^2 r^4/12}{Y\pi r^4/4} = \frac{\pi}{3} \qquad \text{or} \qquad \frac{F_s}{F_c} > 1$$

It shows that the rod with square cross section is more stiff as compared to that with the circular cross section.

8. Cantilever

A beam fixed horizontally at one end and loaded at the other end is known as the cantilever. A cross-section of the cantilever is shown in

Figure 14. Under the effect of an external weight loaded at its free-end Q and due to its own weight, the beam bends as shown in Figure 14.

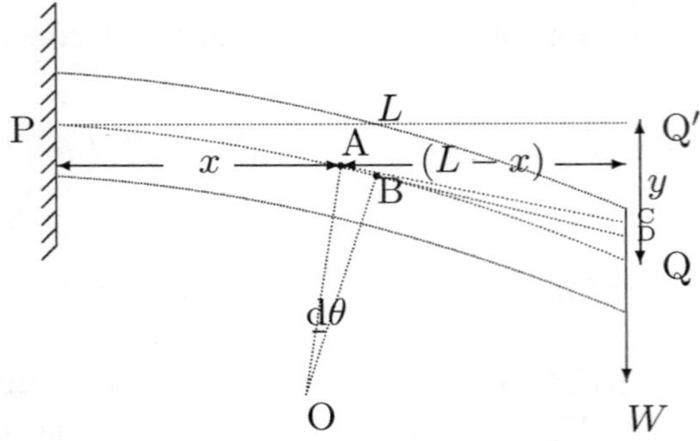

Figure 14: Cross-section of cantilever. W is the external force applied at the free-end Q.

In order to get the expression for the depression of the beam at the free-end, let us make the following assumptions:

1. The area of cross-section of the beam remains unchanged during the process of bending.

2. In the bent position, the radius of curvature of the beam is large as compared to its thickness.

3. Depression of the beam is small as compared to its length.

When the beam is not loaded, let the neutral filament be PQ'. After loading at the free-end, the neutral filament becomes PQ. Suppose the length of the beam is L. In the equilibrium position, the depression of the neutral filament at the free-end is obviously $QQ' = y$. Let A be a point on the neutral surface at a distance x from P. Consider the couple due to external forces at this point. The external forces beyond A are the weight W acting at the free-end Q and the weight of $(L - x)$ length of the beam acting at a distance $(L - x)/2$ from A. Let ω be the force per unit length of the beam, due to its weight, then the force due to the weight of the beam will be $\omega(L - x)$. The moment of these external forces at A is

$$\text{moment of forces} = W(L - x) + \omega(L - x)\frac{(L - x)}{2}$$

$$= W(L - x) + \frac{\omega(L - x)^2}{2}$$

As the beam is in the equilibrium position, the external couple which is trying to bend the beam must be equal to the bending moment of the beam which is trying to restore the beam. Suppose Y is the Young's modulus of the material of the beam, a the area of cross-section, K the radius of gyration and R the radius of curvature of the neutral axis at the point A. The bending moment of the beam at the point A is

$$\text{bending moment of the beam at A} = \frac{YI_g}{R}$$

Thus, in the equilibrium position, we have

$$\frac{YI_g}{R} = W(L - x) + \frac{\omega(L - x)^2}{2} \tag{4.6}$$

This relations shows that when x tends to L, R becomes very large and when x tends to zero, R becomes small. However, for two close points A and B, separated by a small distance $\mathrm{d}x$, the value of R can be taken the same. Let the angle subtended by AB at the center of curvature is $\mathrm{d}\theta$. Now, we have

$$\text{AB} = \mathrm{d}x = R\,\mathrm{d}\theta \qquad \text{or} \qquad R = \frac{\mathrm{d}x}{\mathrm{d}\theta}$$

Putting the value of R in equation (4.6) and on rearranging, we have

$$YI_g\,\mathrm{d}\theta = \left[W(L - x) + \frac{\omega(L - x)^2}{2} \right]\mathrm{d}x \tag{4.7}$$

Let AC and BD are tangents drawn at the points A and B, respectively. Suppose CD is Δy. Then, Δy is the depression of point B with respect to that of the point A. The angle between these two tangents will be $\mathrm{d}\theta$. Thus, we have

$$\text{CD} = \mathrm{d}y = (L - x)\mathrm{d}\theta \tag{4.8}$$

Putting the value of $\mathrm{d}\theta$ from equation (4.8) in (4.7) and on rearranging, we get

$$\mathrm{d}y = \left[\frac{W(L - x)^2}{YI_g} + \frac{\omega(L - x)^3}{2YI_g} \right]\mathrm{d}x \tag{4.9}$$

Total depression of the neutral filament from the horizontal axis passing through the point P can be obtained by integrating equation (4.9) when x varies from 0 to L. Thus, we have

$$y = \frac{W}{YI_g} \int_0^L (L - x)^2\,\mathrm{d}x + \frac{\omega}{2YI_g} \int_0^L (L - x)^3\,\mathrm{d}x$$

$$= \frac{WL^3}{3YI_g} + \frac{\omega L^4}{8YI_g} \tag{4.10}$$

If W_1 is the weight of the beam, then we have $W_1 = \omega L$ or $\omega = W_1/L$. Using the value of ω in equation (4.10), we have

$$y = \frac{WL^3}{3YI_g} + \frac{W_1 L^3}{8YI_g} = \left(W + \frac{3}{8}W_1\right)\frac{L^3}{3YI_g} \tag{4.11}$$

9. Beam supported at the ends and loaded in the middle

Let us consider a beam of uniform area of cross-section and supported at the points P and Q with the help of knife edges and loaded at the middle point C by a weight W, as shown in Figure 15.

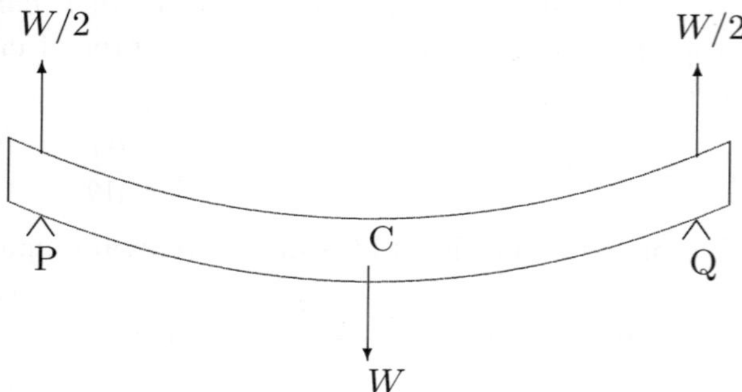

Figure 15: A beam supported at the ends and loaded in the middle.

When the beam is in the equilibrium state, the reaction forces at the edges P and Q are $W/2$ in the upward direction. This arrangement can be considered as equivalent to two inverted cantilevers PC and CQ held horizontal at C and loaded at P and Q, respectively by the weight $W/2$. Let the distance between the knife edges be L. Then, the length of each cantilever is $L/2$. In such a situation, the elevation y of the points P and Q above the point C will be the same as the depression at the ends of cantilever of length $L/2$ and loaded with the weight $W/2$. Suppose W_1 be the weight of the beam. Then weight of half of the beam is $W_1/2$. Using equation 4.11, we have

$$y = \left(\frac{W}{2} + \frac{3}{8}\frac{W_1}{2}\right)\frac{(L/2)^3}{3YI_g} = \left(W + \frac{3}{8}W_1\right)\frac{L^3}{48YI_g}$$

When $W >> W_1$, we have

$$y = \frac{WL^3}{48YI_g}$$

(i) For a beam having circular cross section of radius r, we have area $= \pi r^2$ and

$$I_g = \frac{\pi r^4}{4}$$

Thus, we have

$$y = \frac{4WL^3}{48Y\pi r^4} = \frac{WL^3}{12Y\pi r^4}$$

(ii) For a beam having rectangular cross section of breadth b and thickness d, we have area $= bd$ and

$$I_g = \frac{bd^3}{12}$$

Thus, we have

$$y = \frac{12WL^3}{48Ybd^3} = \frac{WL^3}{4Ybd^3}$$

Exercise 17: A 100 cm long metal bar of mass 1 kg and of square cross section with area 1 cm^2 is supported by two knife edges at its ends. The mid-point of the bar is depressed by 2.5 mm when loaded by a mass of 2 kg. Find out the Young's modulus of the material of the bar.

Solution: The depression y of bar loaded in the middle is

$$y = \left(W + \frac{3}{8}\,W_1\right)\frac{L^3}{48YI_g}$$

Therefore, we have

$$Y = \left(W + \frac{3}{8}\,W_1\right)\frac{L^3}{48yI_g}$$

For a bar, we have breadth $b = $ depth $d = 1$ cm $= 10^{-2}$ m. Therefore,

$$I_g = \frac{bd^3}{12} = \frac{10^{-2} \times (10^{-2})^3}{12} = \frac{10^{-8}}{12}\ \text{m}^3$$

We have $L = 100$ cm $= 1$ m, $y = 2.5$ mm $= 2.5 \times 10^{-3}$ m, $W_1 = 1 \times 9.8$ $= 9.8$ N, $W = 2 \times 9.8 = 19.6$ N. On using the values in the expression, we have

$$Y = \left(19.6 + \frac{3}{8} \, 9.8\right) \frac{1^3}{48 \times 2.5 \times 10^{-3} \times (10^{-8}/12)} = 2.33 \times 10^{11} \text{ N m}^{-2}$$

Exercise 18: A metallic rectangular bar of width 2.1m and thickness 1.2m is supported on two knife edges 1.1 m apart. A load of 2 kg hanging at the center of the bar depresses that point 2.1 mm. Calculate Young's modulus of the material of the bar.

Solution: Length $L = 1.1$ m. Depression $y = 2.1$ mm $= 2.1 \times 10^{-3}$ m. Suspended weight $W = mg = 2 \times 9.8 = 19.6$ N
Geometrical moment of inertia

$$I_g = \frac{bd^3}{12} = \frac{2.1 \times 1.2^3}{12} = 0.3024 \text{ cm}^3 = 0.3024 \times 10^{-6} \text{ m}^3$$

The depression is expressed as

$$y = \frac{WL^3}{48YI_g} \qquad \text{and therefore} \qquad Y = \frac{WL^3}{48yI_g}$$

On using the values, we get

$$Y = \frac{(19.6)1.1^3}{48(2.1 \times 10^{-3})(0.3024 \times 10^{-6})} = 8.558 \times 10^8 \text{ N/m}^2$$

10. Torsion of cylinder

A cylinder is said to be under torsion when its one end (face) is fixed and the other is twisted by a couple acting in a plane perpendicular to this axis. This leads to the twisting of cylinder which however is opposed by a couple, called the restoring couple, developed in the cylinder due to the elasticity of the material of the cylinder. Under equilibrium condition, the restoring couple balances the external couple acting on the cylinder.

　　Let us consider a cylinder of length L and radius r, as shown in Figure 16 (a), where the upper end of the cylinder is fixed and a twisting couple is applied on its lower face in a plane perpendicular to its axis. Suppose η be the modulus of rigidity of the material of the cylinder. As a result of this, a point P, for example, lying on the periphery of its base is moved to a new position P′, as shown in Figure 16 (a).

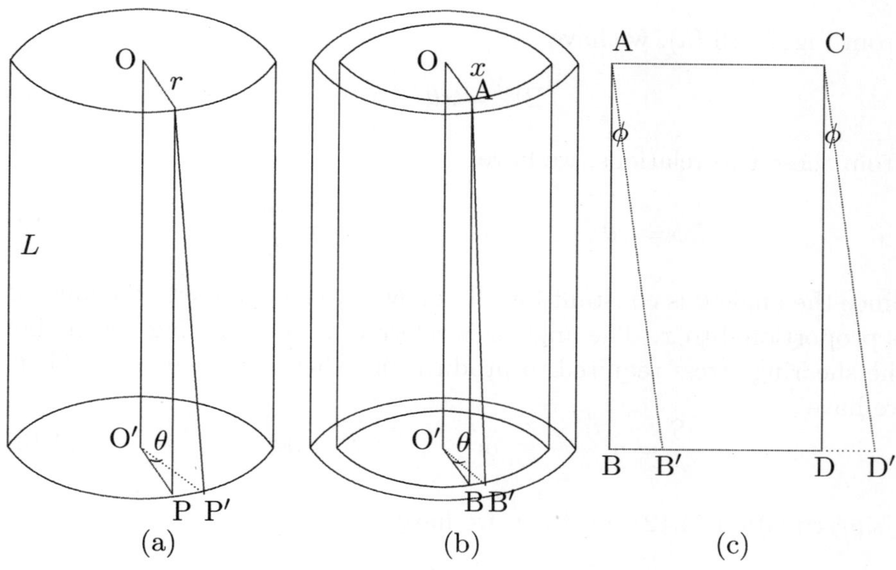

Figure 16: (a) A cylinder of length L and radius r whose upper end is fixed and a couple is applied in the plane of the lower end. (b) Twisting of inner cylinder of radius x. (c) Lamina obtained after cutting and spreading hollow cylinder of radius x.

The arc PP′ subtends an angle θ at the center O′ of the bottom face. This cylinder can be considered as consists of a large number of very thin coaxial cylinders. Let us consider one of these cylinders having radius x, as shown in Figure 16 (b). The point B, for example, on the periphery of this cylinder of radius x moves to the point B′ after twist. The arc BB′ subtends the angle θ at the O′. The angle θ is referred to the angle of twist and is the same for all the coaxial cylinders. In order to calculate the twisting couple, let us consider a thin hollow cylinder of inner radius x and outer radius $x + dx$. The thickness of this cylinder is dx which is very small. Let A be a point on the top of this cylinder. The line AB is parallel to the axis OO′ of the cylinder. If we cut this hollow cylinder along AB and spread it, then we get a lamina ABCD, as shown in Figure 16 (c). The point C coincides with A and the point D coincides with B when the cylinder is intact. When the cylinder is twisted, the point B moves to B′ and the point D moves to D′. The angle through which the cylinder has been sheared is obviously $\angle$ BAB′ $= \angle$ DCD′ $= \phi$. When the angle ϕ is small and expressed in radian, we have

$$\text{BB}' = L\phi$$

From Figure 16 (a), we have

$$\text{BB}' = x\theta$$

From these two relations, we have

$$L\phi = x\theta \qquad \text{or} \qquad \phi = \frac{x\theta}{L} \qquad (4.12)$$

Since the angle θ is constant for all hollow coaxial cylinders, the angle ϕ is proportional to x. The angle ϕ is referred as the shearing strain. Let the shearing stress required to produce this shearing stress is S. Then, we have

$$\frac{S}{\phi} = \eta \qquad \text{or} \qquad S = \eta\phi \qquad (4.13)$$

Using equation (4.12) in (4.13), we have

$$S = \frac{\eta\, x\theta}{L} \qquad (4.14)$$

The face area of the hollow cylinder is

$$\text{face area} = \pi(x + \mathrm{d}x)^2 - \pi x^2 = 2\pi x\, \mathrm{d}x$$

Thus, the shearing force $(\mathrm{d}F)$ acting on the face area of the hollow cylinder is

$$\mathrm{d}F = \text{shearing stress} \times \text{area} = S \times 2\pi x\, \mathrm{d}x \qquad (4.15)$$

Using equation (4.14) in (4.15), we have

$$\mathrm{d}F = \frac{2\pi\eta\, \theta}{L}\, x^2\, \mathrm{d}x$$

The moment $\mathrm{d}\tau$ of force $\mathrm{d}F$ about the axis OO$'$ is

$$\mathrm{d}\tau = \text{force} \times \text{distance} = \frac{2\pi\eta\, \theta}{L}\, x^3\, \mathrm{d}x \qquad (4.16)$$

To get the moment of total external force acting on the face area of the cylinder, $i.e.$, the twisting couple, we have to integrate equation (4.16) for x varying from zero to r. Thus, we have

$$\tau = \int_0^r \frac{2\pi\eta\, \theta}{L}\, x^3\, \mathrm{d}x = \frac{\pi\eta\, \theta r^4}{2L}$$

Thus, the twisting couple per unit twist is

$$C = \frac{\tau}{\theta} = \frac{\pi\eta\, r^4}{2L}$$

The C is referred to the torsional rigidity of the cylinder.

Exercise 19: A wire of 2 mm diameter and 1 m length has one of its ends fixed. Calculate the couple required to produce a twist of $45°$ at the other end when the modulus of rigidity η of the material of wire is 5×10^{10} N m^{-2}.

Solution: The couple τ required to twist the wire is

$$\tau = \frac{\pi\eta\,\theta r^4}{2L}$$

We have $\eta = 5 \times 10^{10}$ N m^{-2}, $\theta = 45° = \pi/4$ rad, $r = 2/2 = 1$ mm $= 10^{-3}$ m, $L = 1$ m. On using the values, we get

$$\tau = \frac{\pi \times 5 \times 10^{10} \times (\pi/4) \times (10^{-3})^4}{2 \times 1} = 0.062\text{N m}$$

Exercise 20: One end of cylindrical wire of 1.3 m length and 3 mm radius is twisted through $50°$. Calculate the angle of shear on its surface.

Solution: We have $l = 1.3$ m and $r = 3$ mm $= 3 \times 10^{-3}$ m. If θ is the angle of twist and ϕ the angle of shear, we have

$$\phi = \frac{r}{l}\,\theta$$

Using the values, we have

$$\phi = \frac{3 \times 10^{-3}}{1.3}\, 50 = 0.115°$$

Exercise 21: A cylindrical wire of length 1.2 m and radius 0.5 mm is clamped at one of its ends. Calculate the couple required to twist the other end by $70°$. Given, the modulus of rigidity $\eta = 2.8 \times 10^{10}$ N/m^2.

Solution: We have $r = 0.5 \times 10^{-3}$ m, $L = 1.2$ m,

$$\theta = 70° = \frac{70\pi}{180}\text{ rad}$$

The couple required is

$$\tau = \frac{\pi\eta\, r^4}{2L}\,\theta = \frac{\pi(2.8 \times 10^{10})(0.5 \times 10^{-3})^4}{2(1.2)}\,\frac{70\pi}{180} = 2.80 \times 10^{-3}\text{ N m}$$

11. Energy stored in a strained body

Elastic bodies can be deformed under application of external forces. This deformation may be a change in length, in volume or in shape. To produce this deformation in a body, work has to be done by the applied force. This amount of energy is stored in the body in the form of its potential energy. This potential energy is often referred to strain energy of the body. On removal of the applied forces, the stress in the body disappears and this stored energy usually converts in the form of heat. As the strains are of three types, it is interesting to see the forms of strained energy in various cases.

11.1 Strain energy in a linearly strained body

Let is consider a wire of length L and of small area of cross-section a. Suppose on application of a tensile force F, the length of the wire increases by l. This increase in length may be considered as a sum of a number of small changes. Let us consider a small change dl. For this change, the amount of work done dW by the force is

$$dW = F\,dl$$

The applied force F and the change in length dl are in the same direction. Thus, the total amount of work done in increasing the length by l is

$$W = \int_0^l F\,dl$$

We have the Young's modulus

$$Y = \frac{\text{normal stress}}{\text{linear strain}} = \frac{F/a}{l/L} = \frac{FL}{al} \qquad \text{or} \qquad F = \frac{alY}{L}$$

Using the expression for F, the amount of work done is

$$W = \int_0^l \frac{alY}{L}\,dl = \frac{aY}{L}\int_0^l dl = \frac{aYl^2}{2L}$$

It can be expressed as

$$W = \frac{1}{2}\,F\,l$$

Hence, the strained energy is

$$\text{strained energy} = \frac{1}{2}\,\text{stretching force} \times \text{change in length}$$

Since the change in length l is very small in comparison to the original length L, we may consider that the volume of the body $V = La$. Thus, the strained energy per unit volume is

$$E = \frac{W}{V} = \frac{Fl/2}{La} = \frac{1}{2} \frac{F}{a} \frac{l}{L}$$

Thus, the strained energy per unit volume is

$$E = \frac{1}{2} \text{ stress} \times \text{linear strain}$$

11.2 Strain energy of the body subjected to volume strain

Let is consider a cube of volume V and whose each face is of area A. Suppose F is the external inward force acing on each face. For convenience, let the sides of the cube are along the x, y, z axes of the Cartesian coordinate system. Due to these inward forces, the volume of the cube will change. Let the change in volume of the cube is v. This change in volume is due to reduction of length of each side of the cube. In order to make a change $\mathrm{d}x$ in the x-direction, the work done is

$$\mathrm{d}W = F\,\mathrm{d}x = \frac{F}{A}\,A\,\mathrm{d}x = P\,\mathrm{d}v$$

where $P = F/A$ is the pressure on each surface and $\mathrm{d}v = A\,\mathrm{d}x$. The P will obviously be the stress on the cube in x-direction. The total work done in changing the volume of the cube by v is

$$W = \int_0^v P\,\mathrm{d}v$$

For the Bulk modulus, we have

$$K = \frac{\text{pressure}}{\text{volume strain}} = \frac{P}{v/V} = \frac{PV}{v} \qquad \text{or} \qquad P = \frac{Kv}{V}$$

Using the expression for P, the amount of work done is

$$W = \int_0^v \frac{Kv}{V}\,\mathrm{d}v = \frac{K}{V}\int_0^v v\,\mathrm{d}v = \frac{K}{V}\frac{v^2}{2}$$

It can be expressed as

$$W = \frac{1}{2}\frac{vK}{V}v = \frac{1}{2}Pv$$

Hence, the strained energy is

$$\text{strained energy} = \frac{1}{2}\text{ pressure} \times \text{change in volume}$$

Since the change in volume v is very small in comparison to the original volume V, we may consider that the volume of the body is V. Thus, the strained energy per unit volume is

$$E = \frac{W}{V} = \frac{Pv/2}{V} = \frac{1}{2} P \frac{v}{V}$$

Thus, the strained energy per unit volume is

$$E = \frac{1}{2} \text{ stress} \times \text{volume strain}$$

11.3 Strain energy of the body subjected to shear strain

Let a cube is sheared by a tangential force on the upper face. Due to this force, the point A on its side-face is shifted to A$'$ and the point B to B$'$, as shown in the following figure.

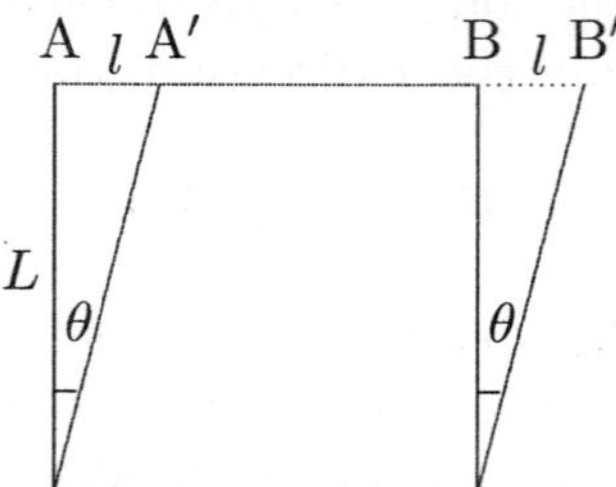

Let the shift in these points is l, so that AA$'$ = BB$'$ = l. This shift may be considered as a sum of a number of small changes. Let us consider a small change dl. For this change, the amount of work done dW by the force is

$$dW = F\, dl$$

The applied force F and the change in the length dl are in the same direction. Thus, the total amount of work done in increasing the length by l is

$$W = \int_0^l F\, dl$$

We have the modulus of rigidity

$$\eta = \frac{\text{tangential stress}}{\text{shear strain}} = \frac{F/a}{\theta} = \frac{F/a}{l/L} = \frac{FL}{al} \qquad \text{or} \qquad F = \frac{al\eta}{L}$$

Using the expression for F, the amount of work done is

$$W = \int_0^l \frac{al\eta}{L}\, dl = \frac{a\eta}{L} \int_0^l l\, dl = \frac{a\eta\, l^2}{2L}$$

It can be expressed as

$$W = \frac{1}{2}\, F\, l$$

Hence, the strained energy is

$$\text{strained energy} = \frac{1}{2}\, \text{tangential force} \times \text{displacement}$$

Since the change in length l is very small in comparison to the original length L, we may consider the volume of the body $V = La$. Thus, the strained energy per unit volume is

$$E = \frac{W}{V} = \frac{Fl/2}{La} = \frac{1}{2} \frac{F}{a} \frac{l}{L}$$

Thus, the strained energy per unit volume is

$$E = \frac{1}{2}\, \text{tangential stress} \times \text{shear strain}$$

12. Searle's method

The Searle's method for determination of elastic constants, Young's modulus Y, modulus of rigidity η, Bulk modulus K and Poisson ratio σ is as the following.

12.1 Young's modulus Y

In the Searle's method, we take two identical metal bars of square or circular cross section and a wire whose elastic constants are to be determined. The ends of the wire are attached in the middle points E and F of the bars, AB and CD, respectively, as shown in Figure 17. The bars AB and CD are suspended at the middle points by two equal parallel threads GH and IJ, respectively, whose other ends are attached to the fixed horizontal platform.

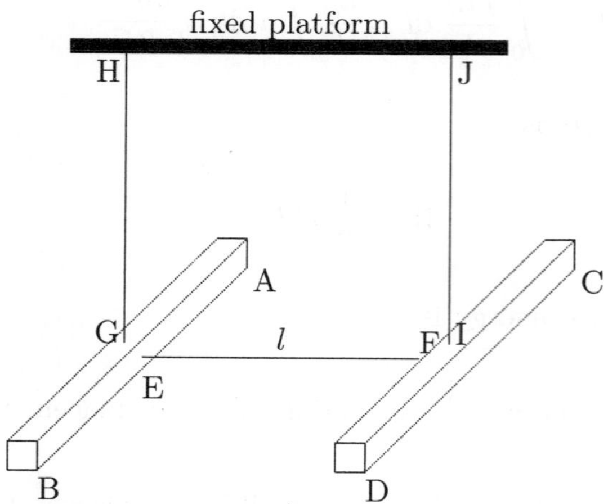

Figure 17: Searle's apparatus for determination of Y.

The ends A and C of the bars are brought near each other symmetrically, as shown in Figure 18, so that the wire EF is bent in the form of a circular arc and then released.

A torque is acted by the wire on the bars and on release, the bars oscillate in a horizontal plane, from the circular arc in one side to a similar arc in the other side. The points E and F remain almost stationary, so that the cation of the wire on the bars and their reaction constitute a couple. If 2θ be the angle subtended by the wire of length l at the center of curvature O of the arc of radius R. We have

$$2\theta = \frac{l}{R} \qquad \text{or} \qquad R = \frac{l}{2\theta}$$

The bending moment M' of the wire is

$$M' = \frac{Y I_g}{R}$$

where I_g is the geometrical moment of inertia. For a wire of radius r, we have $I_g = \pi r^4/4$. Using the values of R and I_g, we have

$$M' = \frac{Y \pi r^4/4}{l/2\theta} = \frac{Y \pi r^4 \theta}{2l}$$

This produces angular acceleration in each of the rods about its suspension and if I be the moment of inertia of a rod about its suspension or an axis passing through its middle and perpendicular to its length, we have

$$I\,\frac{\mathrm{d}^2\theta}{\mathrm{d}t^2} = -\frac{Y\pi r^4\theta}{2l} \qquad \text{or} \qquad \frac{\mathrm{d}^2\theta}{\mathrm{d}t^2} = -\frac{Y\pi r^4\theta}{2lI}$$

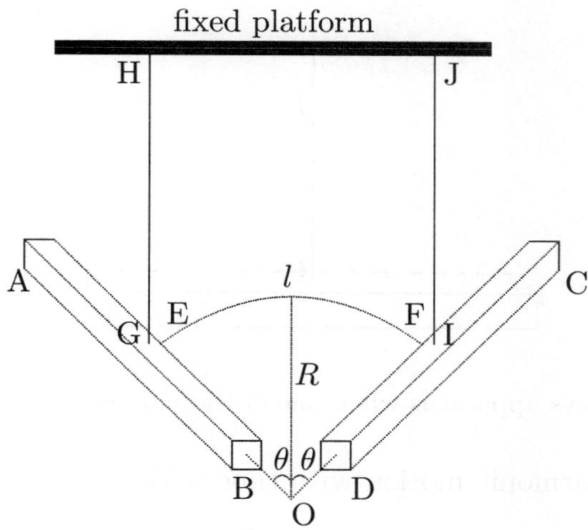

Figure 18: Searle's apparatus when the wire EF is bent.

This is simple harmonic motion with time-period

$$T_1 = 2\pi\sqrt{\frac{2lI}{Y\pi r^4}} \qquad \text{and we have} \qquad Y = \frac{8\pi lI}{r^4 T_1^2} \qquad (4.17)$$

12.2 Modulus of rigidity η

For determination of η, the suspension is removed and one of the bars is clamped horizontally while the other bar hangs vertically below it by the wire EF whose modulus of rigidity η is to be determined.

The suspended bar is turned about the wire EF in a horizontal plane to twist the wire and released. The suspended bar oscillates with a time-period T_2. When the wire is twisted through an angle θ, the restoring couple in it is

$$\frac{\pi\eta r^4\theta}{2l}$$

If I be the moment of inertia of one of the bars about the axis of rotation EF, we have

$$I\,\frac{\mathrm{d}^2\theta}{\mathrm{d}t^2} = \frac{\pi\eta r^4\theta}{2l} \qquad \text{or} \qquad \frac{\mathrm{d}^2\theta}{\mathrm{d}t^2} = \frac{\pi\eta r^4\theta}{2lI}$$

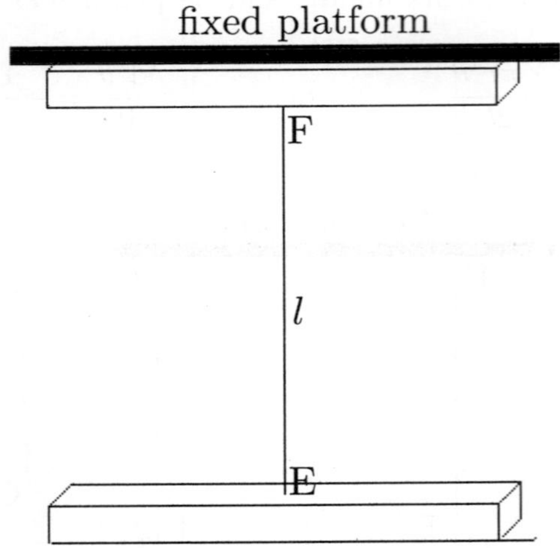

Figure 19: Searle's apparatus when one of the bars is clamped horizontally.

This is simple harmonic motion with time-period

$$T_2 = 2\pi \sqrt{\frac{2lI}{\eta \pi r^4}}$$

Therefore, we have

$$\eta = \frac{8\pi l I}{r^4 T_2^2} \tag{4.18}$$

On dividing equation (4.17) by equation (4.18), we get

$$\frac{Y}{\eta} = \frac{8\pi l I / r^4 T_1^2}{8\pi l I / r^4 T_2^2} = \frac{T_2^2}{T_1^2} \qquad \text{or} \qquad \eta = Y\,\frac{T_1^2}{T_2^2} \tag{4.19}$$

12.3 Poisson ratio σ

From the relation

$$Y = 2\eta(1 + \sigma)$$

we have

$$\sigma = \frac{Y}{2\eta} - 1$$

On using equation (4.20), we have

$$\sigma = \frac{T_2^2}{2T_1^2} - 1 \tag{4.20}$$

12.4 Bulk modulus K

From the relation

$$\frac{9}{Y} = \frac{3}{\eta} + \frac{1}{K}$$

we have

$$K = \frac{\eta Y}{9\eta - 3Y} \tag{4.21}$$

On using the values of Y and η obtained earlier, we can get K with the help of equation (4.21).

13. Multiple choice questions

1. For which of the strain, the stress is not normal.

    (i) Longitudinal strain   (ii) Lateral strain

    (iii) Volume strain   (iv) Shear strain

    Ans. (iv)

2. Which of the following elastic constants is dimensionless.

    (i) Young's modulus Y (ii) Bulk modulus K

    (iii) Modulus of rigidity η (iv) Poisson ratio σ

    Ans. (iv)

3. Which of the following relations between the elastic constants is not correct.

    (i) $Y = 3K(1 + 2\sigma)$ (ii) $Y = 2\eta(1 + \sigma)$

    (iii) $\sigma = \dfrac{3K - 2\eta}{6K + 2\eta}$ (iv) $Y = \dfrac{9K\eta}{3K + \eta}$

    Ans. (i)

4. Which of the following statements in the context of a beam is wrong.

    (i) Material of a beam is elastic.

    (ii) Material of a beam is isotropic.

    (iii) Length and width (or diameter) of a beam are equal.

    (iv) Cross section of a beam is uniform.

    Ans. (iii)

5. Poisson ratio for a material cannot have the value

    (i) 0.7 (ii) 0.2 (iii) 0.1 (iv) 0.5

    Ans. (i)

6. Two wires of the same material and radius have lengths in the ratio 1 : 2. If they are stretched by the same force, then the strain produced in the wires will be in the ratio

    (i) 1 : 1 (ii) 1 : 2 (iii) 2 : 1 (iv) 1 : 4

    Ans. (i)

7. A cylindrical wire of length L and radius r is fixed at one end and a force F is applied on the other end. The extension produced in the wire is l. The extension produced in another wire made up of the same material and of length $2L$ and radius $2r$ by a force $2F$ is

    (i) l (ii) $2l$ (iii) $l/2$ (iv) $4l$

    Ans. (i)

8. A metal beam, having Young's modulus Y, is supported at two ends and is loaded at its center. The depression at the center is proportional to

    (i) Y^2 (ii) Y (iii) $1/Y$ (iv) $1/Y^2$

    Ans. (iii)

9. The upper end of a cylindrical wire 1 m long and 2 mm radius is clamped. The lower end of the wire is twisted through an angle of 45°. Then the angle of shear is

    (i) 0.09° (ii) 0.9° (iii) 9° (iv) 90°

    Ans. (i)

10. For a given material, the Young's modulus is 2.4 times that of modulus of rigidity. Its Poisson ratio is

    (i) 2.4 (ii) 1.2 (iii) 0.4 (iv) 0.2

    Ans. (iv)

11. A wire breaks if it is stretched by more than 3 mm. The wire is cut into two equal parts. Then each part can be stretched without break by

    (i) 0.75 mm (ii) 1.5 mm (iii) 3 mm (iv) 6 mm

    Ans. (ii)

12. A spiral spring is stretched by a weight attached to it. The strain
will be

   (i) volumetric   (ii) tensile   (iii) compressive   (iv) shear

   Ans. (iv)

13. A wire can bear a load of 20 kg without breaking. If the wire is
cut into two equal parts, each part can bear a maximum load of

   (i) 10 kg   (ii) 20 kg   (iii) 40 kg   (iv) 80 kg

   Ans. (ii)

14. Problems and questions

1. Explain the terms, stress and strain. How they express various kinds
of elastic constants.

2. What is Hooke's law. Describe (i) Young's modulus, (ii) Bulk modulus, (iii) Modulus of rigidity and (iv) Poisson ratio.

3. Describe the equivalence of shear strain to compression and extension strains.

4. Derive relations between the Young's modulus, Bulk modulus, Modulus of rigidity and Poisson ratio.

5. Prove that shear stress is equivalent to an equal linear tensile stress
and to an equal linear compressive stress.

6. Explain different kinds of elastic constants. Obtain the relations

$$Y = 3K(1 - 2\sigma) \qquad Y = 2\eta(1 + \sigma)$$

7. Define Poisson ration. Shaw that

$$\sigma = \frac{3K - 2\eta}{6K + 2\eta}$$

8. Define a beam. Obtain expression for bending moment of a beam.

9. A rectangular beam is supported at the ends and loaded in the
middle. Find expression for depression of the beam.

10. What do you mean by torsion of a cylinder. Obtain expression for
the torsional rigidity of the cylinder.

11. Derive relation for twisting couple per unit twist of a cylinder fixed at one end and twisted by a couple at other end.

12. Describe Searle's method for determination of elastic constants.

13. What force would be required to stretch a wire of 2.5 cm^2 cross section, so that its length becomes double of its original length? Given, Young's modulus $Y = 3.6 \times 10^{11}$ N/m^2.

14. A load of 2 kg produces an extension of 1 mm in a wire of 2.6 m length and radius 0.4 mm. Calculate Young's modulus of the material of the wire.

15. A wire of length 3 m has a percentage strain of 0.5% under a tensile force. What is the extension in the wire.

16. Write notes on the following

   (i) Rigid body

   (ii) Stress

   (iii) Strain

   (iv) Normal and tangential stresses

   (v) Longitudinal, lateral, volume and shear strains

   (vi) Poisson ratio

   (vii) Variation of strain with stress and elastic limit

   (viii) Hooke's law

   (ix) Young's modulus

   (x) Bulk modulus

   (xi) Modulus of rigidity

   (xii) Bending moment

   (xiii) Cantilever

   (xiv) Torsion of cylinder

   (xv) Flexural rigidity

V. Waves and Oscillations

In science and technology, we generally deal with two types of waves, known as the (i) electromagnetic waves and (ii) mechanical waves. Electromagnetic radiation, comprising γ-rays, x-rays, UV-radiation, visible spectrum, IR-radiation, microwaves and radio waves, lie under the category of electromagnetic waves. For propagation of electromagnetic waves, the medium is not necessarily required. The largest speed of electromagnetic wave is in the vacuum and is denoted by c and having a value 3×10^8 m/s. The velocity v of electromagnetic radiation in a medium with refractive index n is $v = c/n$. A mechanical wave necessarily requires a medium for its propagation. Example of mechanical wave is the sound wave; wave generated in the surface of water. The waves have in general two modes of propagation, known as the (i) transverse mode and (ii) longitudinal mode. The wave moving under the transverse mode is also known as the transverse wave and the wave moving under the longitudinal mode is also known as the longitudinal wave. In the transverse mode, the motion of particles of the medium is perpendicular to the direction of propagation of the wave whereas in the longitudinal mode, the motion of particles of the medium is along the direction of propagation of the wave. Electromagnetic wave has only transverse mode whereas a mechanical wave can have transverse or longitudinal mode. In this chapter, we shall discuss about various types of waves and oscillations.

1. Motion

When position of a body (object) in the space changes with time, the body is said to be in motion. When the position of a body changes along a straight line, the motion is said to be translational. Example of such motion is the motion of a train on a straight track on the surface of the earth; falling of a body from a height under the gravitational pull. When the position of a body changes along a circular arc, the motion is said to be circular.

1.1 Harmonic motion

When the position of a body repeats after a regular interval of time, the motion is said to be harmonic or oscillatory. Example of such motion may be the motion of a body moving along a circular path with constant speed. For a harmonic motion, we define as the following.

A. Cycle

When a body executing harmonic motion reaches again at the same position and in the same phase, the body is said to complete one cycle.

B. Time-period

The time-period of a harmonic motion is the time required a complete one cycle. It is generally denoted by T and is expressed in second.

C. Frequency

The frequency of a harmonic motion is the number of cycles completed in one second. It is generally denoted by ν and is expressed in per second or Hertz. Obviously, the relation between the time-period T and frequency n is

$$n = \frac{1}{T} \qquad \text{or} \qquad T = \frac{1}{n}$$

1.2 Simple harmonic motion

When the harmonic motion of a body is along a straight line, the motion is said to be simple harmonic motion. Example of such motion may be motion of a pendulum with small amplitude under the gravitational force; vibrational motion of a diatomic molecule. In a simple harmonic motion, the force acting on the moving body is always directed towards the equilibrium (mean) position of the motion and is proportional to the distance from the equilibrium position. This force is generally known as the restoring force, as it is always towards the equilibrium position.

D. Amplitude

In case of simple harmonic motion, we have one more parameter, known as the amplitude. It is the maximum distance from the equilibrium position moved by the body under simple harmonic motion. If we denote it by A, then the body can move up to a distance A on both sides of the equilibrium position.

Exercise 1: If frequency of a source of simple harmonic wave is 400 Hz and the velocity of wave in the medium is 320 m/s. Calculate the distance traveled the wave in the time for 30 vibrations made by the source.

Solution: We have frequency $n = 400$ Hz and velocity of the wave $v = 32000$ cm/s. The wavelength λ of the wave is

$$\lambda = \frac{v}{n} = \frac{32000}{400} = 80 \text{ cm}$$

The distance traveled by the wave in one vibration is 80 cm. Therefore, the distance traveled by the wave in 30 vibrations is $80 \times 30 = 2400$ cm

2. Mathematical treatment of simple harmonic motion

Let us consider a particle of mass m executing simple harmonic motion. The motion is along a straight line. Suppose at any instant t, the distance of the particle from the mean (equilibrium) position is x. The restoring force F acting on the particle will be proportional to the distance x and we have

$$F \propto -x \qquad \text{or} \qquad F = -kx$$

Here, the negative sign shows that the restoring force is always towards the equilibrium position and the proportionality constant k is known as the force constant. When $\mathrm{d}^2x/\mathrm{d}t^2$ is the acceleration of particle of mass m, we have

$$m\frac{\mathrm{d}^2x}{\mathrm{d}t^2} = -kx \qquad \text{or} \qquad \frac{\mathrm{d}^2x}{\mathrm{d}t^2} + \omega^2 x = 0 \qquad (5.1)$$

where $\omega = \sqrt{k/m}$ is known as the angular frequency. Equation (5.1) is known as the differential equation of simple harmonic motion (S.H.M.). In order to get solution of equation (5.1), let us assume

$$x = C\,e^{\alpha t}$$

where C and α are constants. On differentiating, we have

$$\frac{\mathrm{d}x}{\mathrm{d}t} = C\alpha\,e^{\alpha t} \qquad \text{and} \qquad \frac{\mathrm{d}^2x}{\mathrm{d}t^2} = C\alpha^2\,e^{\alpha t} \qquad (5.2)$$

Using equations (5.2) in (5.1), we get

$$C\,e^{\alpha t}(\alpha^2 + \omega^2) = 0$$

Since $C\,\mathrm{e}^{\alpha t} \neq 0$, we have

$$\alpha^2 + \omega^2 = 0 \qquad\qquad \text{or} \qquad\qquad \alpha = \pm i\omega$$

where $i = \sqrt{-1}$ is the imaginary number. Since α has two values, possible solutions of equation (5.1) are

$$x = C\,\mathrm{e}^{i\omega t} \qquad\qquad \text{and} \qquad\qquad x = C\,\mathrm{e}^{-i\omega t}$$

A general solution of equation (5.1) is a linear combination of these two solutions and we have

$$x = C_1\,\mathrm{e}^{i\omega t} + C_2\,\mathrm{e}^{-i\omega t}$$

where C_1 and C_2 are arbitrary constants. This equation can be expressed as

$$x = C_1(\cos \omega t + i\sin \omega t) + C_2(\cos \omega t - i\sin \omega t)$$

$$= (C_1 + C_2)\cos \omega t + i(C_1 - C_2)\sin \omega t$$

Using $(C_1 + C_2) = A\sin \delta$ and $i(C_1 - C_2) = A\cos \delta$, we have

$$x = A\sin \delta\ \cos \omega t + A\cos \delta\ \sin \omega t = A\sin(\omega t + \delta) \qquad (5.3)$$

This is the required solution of equation (5.1) and gives the displacement of the particle executing S.H.M. at any time t. Here, A is the maximum displacement of the particle from its equilibrium position. The quantity $(\omega t + \delta)$ is known as the phase of the motion of the particle. At the time $t = 0$, the phase δ is known as the initial phase of the motion. When the particle starts motion from its mean position, the value of δ is zero. But, when the particle starts motion from any of the extreme positions, the value of δ is $\pi/2$. Let us see what happens when we replace t by $(t + 2\pi/\omega)$ in equation (5.3). We have

$$x = A\sin(\omega[t + 2\pi/\omega] + \delta)$$

$$= A\sin(\omega t + 2\pi + \delta) = A\sin(\omega t + \delta) \qquad (5.4)$$

It shows that the motion of the particle is repeated after a time-interval of $(2\pi/\omega)$. Thus, the time-period T of the motion is

$$T = \frac{2\pi}{\omega}$$

Using here the value of ω, we have the time-period

$$T = 2\pi\sqrt{\frac{m}{k}}$$

The frequency n of the S.H.M. is

$$n = \frac{1}{T} = \frac{1}{2\pi}\sqrt{\frac{k}{m}}$$

Exercise 2: The displacement x of a particle executing S.H.M. is expressed as $x = A\sin(\omega t + \delta)$. Show that the time-period of the motion is $(2\pi/\omega)$.

Solution: We know that the time-period of S.H.M. is the time required a complete one cycle. The displacement x of particle from its mean position at time t is

$$x = A\sin(\omega t + \delta)$$

After a time-interval $(2\pi/\omega)$, the time becomes $t + (2\pi/\omega)$. At this time, the displacement is

$$x' = A\sin(\omega[t + 2\pi/\omega] + \delta)$$

$$= A\sin(\omega t + 2\pi + \delta) = A\sin(\omega t + \delta)$$

This $x' = x$ shows that after a time-interval of $(2\pi/\omega)$, the particle is at the same position. Thus, the time-period of the S.H.M. is $(2\pi/\omega)$.

2.1 Velocity and acceleration of the particle

Equation (5.4) for displacement of the particle executing S.H.M. can be used for finding out velocity and acceleration of the particle. The velocity v of the particle is

$$v = \frac{dx}{dt} = A\omega\cos(\omega t + \delta) = A\omega\sqrt{1 - \sin^2(\omega t + \delta)}$$

Using here the value of $\sin(\omega t + \delta)$ from equation (5.4), we get

$$v = \omega\sqrt{A^2 - x^2}$$

This expression gives the velocity v of the particle at a position x. It shows that the velocity is the maximum at the mean position $(x = 0)$ and is

$$v_m = A\omega = \frac{A2\pi}{T}$$

Further, the velocity v is zero when the particle is at the extreme ($x = A$). The acceleration a of the particle is

$$a = \frac{d^2x}{dt^2} = -A\omega^2 \sin(\omega t + \delta) = -\omega^2 x$$

This expression gives the acceleration a of the particle at a position x. It shows that the acceleration at the mean position ($x = 0$) is zero. Further, the acceleration is the maximum at each of the extremes ($x = A$). That is

$$|a_m| = \omega^2 A$$

Exercise 3: A particle executes simple harmonic motion along a line has velocity 16 cm/s and 12 cm/s when is at a distance 3 cm and 4 cm, respectively, from the mean position. Calculate (i) the amplitude and (ii) time-period of the oscillation.

Solution: When the particle is at a distance x from the mean position, its velocity is

$$v = \omega\sqrt{A^2 - x^2}$$

On using the values, we have

$$16 = \omega\sqrt{A^2 - 3^2} \qquad \text{and} \qquad 12 = \omega\sqrt{A^2 - 4^2}$$

On diving, we get

$$\frac{4}{3} = \frac{\sqrt{A^2 - 9}}{\sqrt{A^2 - 16}} \qquad \text{or} \qquad \frac{16}{9} = \frac{A^2 - 9}{A^2 - 16} \qquad \text{or} \qquad A = 5 \text{ cm}$$

Using the value of A, we have

$$16 = \omega\sqrt{5^2 - 3^2} \qquad \text{or} \qquad \omega = 4 \text{ s}^{-1}$$

Exercise 4: A particle executes simple harmonic motion in a line 8 cm long. Its velocity when passing through the center of the line is 16 cm/s. Find time-period of motion.

Solution: The amplitude of the motion is $A = 8/2 = 4$ cm. For the angular frequency ω, the velocity v_m at the mean position is

$$v_m = A\omega \qquad \text{or} \qquad \omega = \frac{v_m}{A}$$

Using the values, we have

$$\omega = \frac{16}{4} = 4 \text{ s}^{-1}$$

Thus, the time-period T is

$$T = \frac{2\pi}{\omega} = \frac{2\pi}{4} = \frac{\pi}{2} \text{ s}$$

Exercise 5: A particle executes simple harmonic motion has a time-period of 0.2 s and amplitude 6 cm. Calculate its acceleration when the particle is 0.4 cm away from its mean position and its maximum velocity.

Solution: For the time-period $T = 0.2$ s, the angular frequency ω is

$$\omega = \frac{2\pi}{T} = \frac{2\pi}{0.2} = 10\pi \text{ s}^{-1}$$

The acceleration is

$$a = \omega^2 x = (10\pi)^2 0.4 = 25\pi^2 \text{ cm/s}^2$$

The maximum velocity is

$$v_m = \omega A = 10\pi 6 = 60\pi \text{ cm/s}$$

2.2 Energy of harmonic oscillator

The energy of a harmonic oscillator is the sum of its kinetic energy and the potential energy. For one dimensional motion, the relation between the acting force F and the potential energy V is expressed as

$$F = -\frac{dV}{dx}$$

The potential energy is thus

$$V = -\int F \, dx = k \int x \, dx$$

Here, we have used the expression for F. Hence, we have

$$V = \frac{1}{2} kx^2 + C$$

where C is the constant of integration. At the mean position, the restoring force is zero and therefore, at $x = 0$, the potential energy is zero. It gives the constant of integration $C = 0$. The the potential energy of the particle is therefore

$$V = \frac{1}{2}\, kx^2$$

It shows that the potential energy is the maximum at the extremes of the position. It changes from the maximum value to zero when the particle moves from any of the extreme positions to the mean position. Using the expressions for the force constant k and the displacement x, we have

$$V = \frac{1}{2}\, m\omega^2 A^2 \sin^2(\omega t + \delta) \qquad (5.5)$$

The kinetic energy of the particle is

$$K = \frac{1}{2}\, mv^2$$

Using the expression for v, we have

$$K = \frac{1}{2}\, m\omega^2 A^2 \cos^2(\omega t + \delta) \qquad (5.6)$$

The kinetic energy can also be expressed as

$$K = \frac{1}{2}\, m\omega^2 [A^2 - A^2 \sin^2(\omega t + \delta)] = \frac{1}{2}\, m\omega^2 (A^2 - x^2)$$

It shows that the kinetic energy is the maximum at the mean position and it becomes zero at the extremes. The total energy E is the particle is

$$E = K + V$$

Using equations (5.5) and (5.6), we get

$$E = \frac{1}{2}\, m\omega^2 A^2 \cos^2(\omega t + \delta) + \frac{1}{2}\, m\omega^2 A^2 \sin^2(\omega t + \delta) = \frac{1}{2}\, m\omega^2 A^2$$

Using $T = 2\pi/\omega$, total energy can be expressed as

$$E = \frac{2m\pi^2 A^2}{T^2}$$

Since total energy of the particle does not depend on its position, the energy E of the particle executing S.H.M. is constant. It shows that during the S.H.M., the kinetic energy changes into the potential energy and vice-versa.

Exercise 6: The displacement x of a particle of mass m executing S.H.M. with force constant k is expressed as $x = A\sin(\omega t + \delta)$. Show that average potential energy of the particle is equal to its average kinetic energy.

Solution: The potential energy V and kinetic energy K of the particle executing S.H.M. are

$$V = \frac{1}{2}\, m\omega^2 A^2 \sin^2(\omega t + \delta) \qquad \text{and} \qquad K = \frac{1}{2}\, m\omega^2 A^2 \cos^2(\omega t + \delta)$$

Obviously, the potential and kinetic energies vary with time and they are never negative. We can compute average of them over a time-period $T = 2\pi/\omega$. The average potential energy over the time-period is

$$\langle V \rangle = \frac{1}{T} \int_0^T V \, dt$$

Using expression for V, we get

$$\langle V \rangle = \frac{1}{2T} \int_0^T m\omega^2 A^2 \sin^2(\omega t + \delta) \, dt$$

$$= \frac{m\omega^2 A^2}{4T} \int_0^T [1 - \cos 2(\omega t + \delta)] \, dt$$

$$= \frac{m\omega^2 A^2}{4T} \left[t - \frac{\sin 2(\omega t + \delta)}{2\omega} \right]_0^T$$

$$= \frac{m\omega^2 A^2}{4T} \left[T - \frac{\sin 2(\omega T + \delta)}{2\omega} + \frac{\sin 2\delta}{2\omega} \right] = \frac{m\omega^2 A^2}{4}$$

Here, we used $T = 2\pi/\omega$. The average kinetic energy of the particle is

$$\langle K \rangle = \frac{1}{T} \int_0^T K \, dt$$

Using expression for K, we get

$$\langle K \rangle = \frac{1}{2T} \int_0^T m\omega^2 A^2 \cos^2(\omega t + \delta) \, dt$$

$$= \frac{m\omega^2 A^2}{4T} \int_0^T [1 + \cos 2(\omega t + \delta)] \, dt$$

$$= \frac{m\omega^2 A^2}{4T}\left[t + \frac{\sin 2(\omega t + \delta)}{2\omega}\right]_0^T$$

$$= \frac{m\omega^2 A^2}{4T}\left[T + \frac{\sin 2(\omega T + \delta)}{2\omega} - \frac{\sin 2\delta}{2\omega}\right] = \frac{m\omega^2 A^2}{4}$$

Here, we used $T = 2\pi/\omega$. It shows that the average potential energy of a particle executing S.H.M. is equal to the average kinetic energy of the particle.

Exercise 7: A particle of mass 100 g is executing S.H.M. with time-period of 2π seconds. The displacement of the particle from its equilibrium position is 80 cm at time $\pi/4$ second. Calculate amplitude of oscillations and total energy of the particle.

Solution: The time period $T = 2\pi$ s. Mass $m = 100$ g $= 0.1$ kg. The displacement $x = 80$ cm $= 0.8$ m at time $t = \pi/4$ s. For S.H.M., the displacement is

$$x = A\sin(\omega t) = A\sin\left(\frac{2\pi}{T}\, t\right)$$

Thus, we have

$$0.8 = A\sin\left(\frac{2\pi}{2\pi}\frac{\pi}{4}\right) = \frac{A}{\sqrt{2}} \qquad \text{or} \qquad A = 0.8\sqrt{2}$$

The amplitude of oscillations is $A = 0.8\sqrt{2} = 1.13$ m. Total energy of the particle is

$$E = \frac{2m\pi^2 A^2}{T^2} = \frac{2 \times 0.1 \times \pi^2 \times 0.8^2 \times 2}{4\pi^2} = 0.064 \text{ J}$$

Exercise 8: A particle executing S.H.M. with time-period of 2π s has total energy 9×10^{-3} J. The displacement of the particle from its equilibrium position is 60 cm at time $\pi/4$ second. Calculate the amplitude of oscillations and mass of the particle.

Solution: The time period $T = 2\pi$ s. Total energy $E = 9 \times 10^{-3}$ J The displacement $x = 60$ cm $= 0.6$ m at time $t = \pi/4$ s. For S.H.M., the displacement is

$$x = A\sin(\omega t) = A\sin\left(\frac{2\pi}{T}\, t\right)$$

Thus, we have

$$0.6 = A\sin\left(\frac{2\pi}{2\pi}\frac{\pi}{4}\right) = \frac{A}{\sqrt{2}} \qquad \text{or} \qquad A = 0.6\sqrt{2}$$

The amplitude of oscillations is $A = 0.6\sqrt{2} = 0.85$ m. Mass of the particle is

$$m = \frac{ET^2}{2\pi^2 A^2} = \frac{9 \times 10^{-3} \times 4\pi^2}{2\pi^2 \times 0.6^2 \times 2} = 2.5 \times 10^{-2} \text{ kg} = 25 \text{ g}$$

Exercise 9: A particle is executing S.H.M. with time-period of 30 s and amplitude 15 cm. Calculate the angular frequency of motion and the value of maximum velocity of the particle.

Solution: We have time-period $T = 30$ s and amplitude $A = 15$ cm $= 0.15$ m. Angular frequency of the particle is

$$\omega = \frac{2\pi}{T} = \frac{2 \times 3.14}{30} = 0.209 \text{ rad/s}$$

Maximum velocity v_m of the particle is

$$v_m = A\omega = 0.15 \times \frac{2 \times 3.14}{30} = 0.0314 \text{ m/s}$$

Exercise 10: A particle of mass 1 kg, attached to a mass-less spring, is allowed to execute S.H.M. with an amplitude of 2 m. During the motion, when the particle is at the maximum displacement, its total energy is 8 J. Calculate the force constant of the spring and time-period of motion.

Solution: Amplitude of oscillations $A = 2$ m. Total energy $E = 8$ J. We therefore have

$$E = \frac{1}{2} kA^2 \qquad \text{or} \qquad 8 = \frac{1}{2} k \times 2^2 \qquad \text{or} \qquad k = 4 \text{ J/m}$$

The time-period

$$T = 2\pi \sqrt{\frac{m}{k}} = 2\pi \sqrt{\frac{1}{4}} = \pi = 3.14 \text{ s}$$

Exercise 11: A particle is executing S.H.M. of amplitude 80 cm and time-period 6 s. Find out the time taken by the particle in moving from one end of its path to a position 40 cm away from the mean position on the same side.

Solution: We have amplitude $A = 80$ cm $= 0.8$ m, time-period $T = 6$ s and displacement $x = 40$ cm $= 0.4$ m. The displacement of a particle executing S.H.M. is

$$x = A \sin(\omega t + \delta)$$

Let initially the particle be at one end so that at $t = 0$, we have $x = A$. Then we have

$$A = A \sin(\delta) \qquad \text{or} \qquad \sin(\delta) = 1 = \sin \pi/2$$

Thus, we have $\delta = \pi/2$. Now, the displacement of a particle is

$$x = A \sin(\omega t + \pi/2) = A \cos \omega t = A \cos\left(\frac{2\pi}{T} t\right)$$

Using values of x and T here, we have

$$0.4 = 0.8 \cos\left(\frac{2\pi}{6} t\right) \qquad \text{or} \qquad \cos\left(\frac{\pi}{3} t\right) = \frac{1}{2} = \cos\left(\frac{\pi}{3}\right)$$

Therefore, we have

$$\frac{\pi}{3} t = \frac{\pi}{3} \qquad \text{or} \qquad t = 1 \text{ s}$$

Exercise 12: Find the maximum velocity and maximum acceleration of a particle executing S.H.M. of time-period 4π s and amplitude 2 m.

Solution: We have amplitude $A = 2$ m and time-period $T = 4\pi$ s. The angular frequency is

$$\omega = \frac{2\pi}{T} = \frac{2\pi}{4\pi} = 0.5 \text{ rad/s}$$

The maximum velocity is

$$v_m = A\omega = 2 \times 0.5 = 1 \text{ m/s}$$

The magnitude of maximum acceleration is

$$a_m = A\omega^2 = 2 \times 0.5^2 = 0.5 \text{ m/s}^2$$

3. Simple pendulum

An ideal simple pendulum consists of a heavy point mass suspended by a mass-less, inextensible and perfectly flexible string from a rigid support. In practice, such requirements cannot be fulfilled. Thus, a simple pendulum consists of a small heavy metallic sphere (generally called the bob) suspended by a long fine string from a rigid support. The arrangement of simple pendulum is shown in Figure where the equilibrium position of the pendulum is AO. When the bob is displaced from its equilibrium position and released, its oscillates to and fro about its equilibrium position. The motion of the mass is said to be S.H.M. Suppose at any instant t, is at the point P which is at a distance x from its equilibrium position and the string makes a small angle α about the direction AO. The length of the string is l.

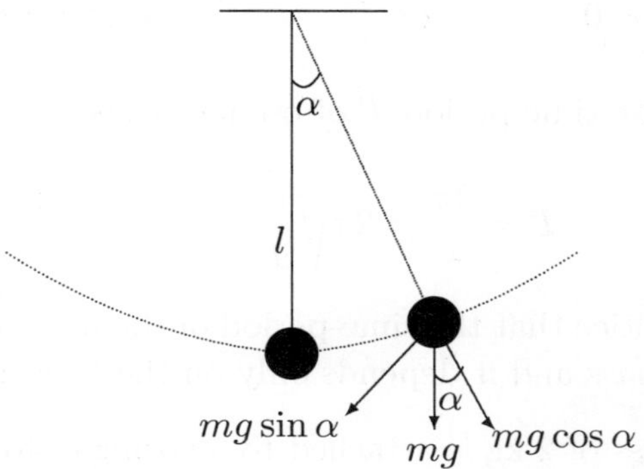

Figure 1: Simple pendulum

At point P, the particle is under the action of two forces: (i) the weight mg of the bob acting vertically downward and (ii) the tension T in the string along PA. The weight mg can be resolved into two components: (a) the radial component $mg \cos \alpha$ which is working in the direction opposite to T and (b) the transverse component $mg \sin \alpha$ working perpendicular to the direction PA. As the string is inextensible, we have $T = mg \cos \alpha$. The force $mg \sin \alpha$ tends to bring the bob to its mean (equilibrium) position. Hence, $mg \sin \alpha$ is the restoring force and we have

$$F = -mg \sin \alpha \qquad (5.7)$$

The negative sign here indicates that the acceleration and displacement are oppositely directed. If x be the displacement at any time, the acceleration is d^2x/dt^2 and force is

$$F = m \, \frac{d^2x}{dt^2} \qquad (5.8)$$

From equations (5.7) and (5.8), we have

$$m \, \frac{d^2x}{dt^2} = -mg \sin \alpha \qquad \text{or} \qquad \frac{d^2x}{dt^2} = -g \sin \alpha \qquad (5.9)$$

For small angle α, we have[1]

$$\sin \alpha = \alpha = \frac{x}{l} \tag{5.10}$$

Using equation (5.10) in (5.9), we get

$$\frac{\mathrm{d}^2 x}{\mathrm{d}t^2} + \frac{g}{l}\, x = 0 \qquad \text{or} \qquad \frac{\mathrm{d}^2 x}{\mathrm{d}t^2} + \omega^2 x = 0 \tag{5.11}$$

where $\omega = \sqrt{g/l}$. The time-period T of oscillation is

$$T = \frac{2\pi}{\omega} = 2\pi\sqrt{\frac{l}{g}}$$

It is interesting to notice that the time-period of the motion of the bob is independent of its mass and it depends only on the length of the string.

Exercise 13: A mass of 2 kg is attached to a spring of force constant 8 N/m. The system is allowed to make S.H.M. Find the natural frequency of oscillations.

Solution: We have mass $m = 2$ kg and force constant $k = 8$ N/m. The natural frequency of oscillations is

$$n = \frac{1}{2\pi}\sqrt{\frac{k}{m}} = \frac{1}{2\pi}\sqrt{\frac{8}{2}} = \frac{1}{\pi}\ \text{Hz}$$

Exercise 14: A simple pendulum of 1.2 m length, hanging vertically at one end, is set to make S.H.M. Calculate the time-period of the oscillations of the pendulum.

Solution: We have length of pendulum $l = 1.2$ m and acceleration due to gravity $g = 9.8$ m/s^2. The time-period of oscillations is

$$T = 2\pi\sqrt{\frac{l}{g}} = 2 \times 3.14\sqrt{\frac{1.2}{9.8}} = 2.2\ \text{s}$$

Exercise 15: Time-period of a simple pendulum 4 s. Calculate the time-period of the pendulum when its length is made four times that of its original length.

---

[1]We have

$$\sin \alpha = \alpha + \frac{\alpha^3}{3!} + \frac{\alpha^5}{5!} + \cdots$$

For small value of α, we can neglect the terms having higher powers of α in comparison to α and thus, we have $\sin \alpha = \alpha$.

Solution: We have time-period $T = 4$ s. Let the length of the pendulum be l. The we have

$$T = 2\pi\sqrt{\frac{l}{g}} \qquad \text{or} \qquad 4 = 2\pi\sqrt{\frac{l}{g}}$$

Let T' be the time-period of oscillations when the length of the pendulum is increased to $4l$. Then we have

$$T' = 2\pi\sqrt{\frac{4l}{g}}$$

From these equations, we have

$$\frac{T'}{4} = \frac{\sqrt{4l/g}}{\sqrt{l/g}} = 2$$

Thus, we have $T' = 8$ s.

Exercise 16: A simple pendulum of length 1 m is hanging in a lift which is ascending at an acceleration of 2.2 m/s^2. Calculate the period of oscillation of the pendulum.

Solution: We have acceleration $f = 2.2$ m/s^2 and length $l = 1$ m. The lift is ascending at an acceleration of 3.2 m/s^2. The acceleration due to gravity is 9.8 m/s^2. Thus, total acceleration is $9.8 + 2.2 = 12$ m/s^2. The time-period of oscillations is

$$T = 2\pi\sqrt{\frac{l}{g}} = 2\pi\sqrt{\frac{1}{12}} = 1.81 \text{ s}$$

4. Mass-string system

When a spring is compressed or stretched by applying a force on it, for small displacements, the elastic force (restoring force) is proportional to the displacement from the mean (equilibrium) position. Therefore,

$$F \propto -x \qquad \text{or} \qquad F = -kx$$

Here, the negative sign shows that the restoring force is always towards the equilibrium position and the proportionality constant k is known as the force constant or stiffness constant of spring. We can have two cases: (i) horizontal oscillations and (ii) vertical oscillations.

4.1 Horizontal oscillations

Let us consider a mass-less string lying in a horizontal plane and one end of which is connected to a point mass m and the other end is connected to a fixed point as shown in Figure 2. The mass m is free to move on a frictionless horizontal surface. When the mass is pulled to the right through a small distance x, then the restoring force exerted by the spring is directed towards the left and can be expressed as

$$F = -kx \tag{5.12}$$

Here, k is known as the force constant or stiffness constant of spring.

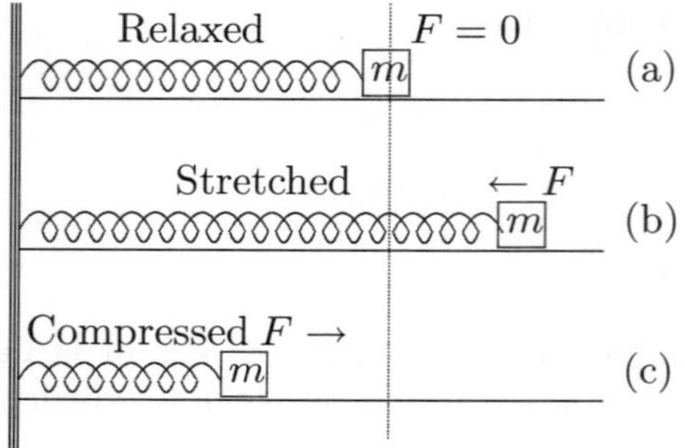

Figure 2: A body of mass m oscillates on a horizontal plane. In the position (a), no force acts on the spring; in position (b), the spring is stretched and the force acts in the left direction; in position (c), the spring is compressed and the force acts in the right direction.

Here, the negative sign shows that restoring force and displacement are oppositely directed. The mass moves with a linear acceleration $\mathrm{d}^2x/\mathrm{d}t^2$ and the force is

$$F = m\,\frac{\mathrm{d}^2x}{\mathrm{d}t^2} \tag{5.13}$$

From equations (5.12) and (5.13), we have

$$m\,\frac{\mathrm{d}^2x}{\mathrm{d}t^2} = -kx \qquad \text{or} \qquad \frac{\mathrm{d}^2x}{\mathrm{d}t^2} + \omega^2 x = 0$$

where $\omega = \sqrt{k/m}$ is generally known as the angular frequency. It is also known as the natural frequency of the oscillator. Thus, the time-period T of oscillation is

$$T = \frac{2\pi}{\omega} = 2\pi\sqrt{\frac{m}{k}}$$

The time period of oscillations depend on the mass m and the force constant k.

4.2 Vertical oscillations

Let us consider a perfectly elastic and mass-less spring of length l is hanging freely from a support (Figure 3). When a point-mass m is attached to its lower end, the spring is stretched through some length due to gravitational force mg acting on the string in the downward direction. This situation corresponds to the equilibrium position of the system. Now, we pull down the mass through a small distance from the equilibrium position and then release. The mass starts oscillating to and fro about the equilibrium position. Suppose at any instant t, the displacement of the mass from the equilibrium position is x, the restoring force exerted by the spring in the upward direction can be expressed as

$$F = -kx \qquad (5.14)$$

where k is known as the force constant or stiffness constant of spring.

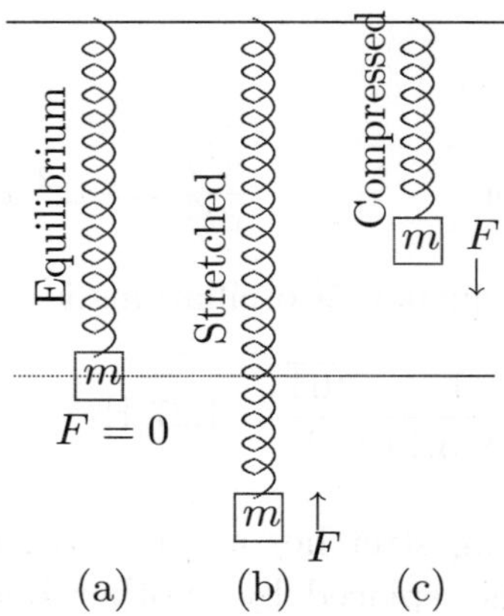

Figure 3: A body of mass m oscillates on a vertical plane. In the position (a), the spring is in the equilibrium position; in position (b), the spring is stretched and the force acts in the up direction; in position (c), the spring is compressed and the force acts in the down direction.

where k is the force constant or stiffness constant of spring. Here, the negative sign shows that restoring force and displacement are oppositely directed. The mass moves with a linear acceleration $\mathrm{d}^2x/\mathrm{d}t^2$ and the force is

$$F = m\,\frac{\mathrm{d}^2x}{\mathrm{d}t^2} \qquad (5.15)$$

Thus, from equations (5.14) and (5.15), we have

$$m\,\frac{\mathrm{d}^2x}{\mathrm{d}t^2} = -kx \qquad \text{or} \qquad \frac{\mathrm{d}^2x}{\mathrm{d}t^2} + \omega^2 x = 0$$

where $\omega = \sqrt{k/m}$ is generally known as the angular frequency. It is also known as the natural frequency of the oscillator. Thus, the time-period T of oscillations is

$$T = \frac{2\pi}{\omega} = 2\pi\sqrt{\frac{m}{k}}$$

The time period of oscillations depend on the mass m and the force constant k.

Exercise 17: A particle of mass 1 kg placed in a potential field $U = 32x^2 + 5$ J/kg is set to execute S.H.M. Calculate the natural frequency of oscillations.

Solution: The potential field is $U = 32x^2 + 5$ and mass $m = 1$ kg. The force due to the potential field acting on the particle is

$$F = -\frac{dU}{dx} = -64x$$

For a particle of mass m, we have

$$m\,\frac{d^2x}{dt^2} = -64x \qquad \text{or} \qquad \frac{d^2x}{dt^2} = -\omega^2\,x$$

where $\omega^2 = 64/m$. The natural frequency of oscillations is

$$n = \frac{\omega}{2\pi} = \frac{1}{2\pi}\sqrt{\frac{64}{m}} = \frac{1}{2 \times 3.14}\sqrt{\frac{64}{1}} = 1.27 \text{ Hz}$$

Exercise 18: A body of mass 4 kg stretches a spring 0.2 m from its equilibrium position. The mass is replaced by another body of 2 kg and the system is set to make S.H.M. Calculate the time-period and frequency of oscillations.

Solution: For the body of mass 4 kg, we have $x = 0.2$ m. Thus, we have

$$F = kx \qquad \text{or} \qquad k = \frac{F}{x} = \frac{mg}{x} = \frac{4 \times 9.8}{0.2} = 196 \text{ N/m}$$

For the body of mass 2 kg, the time-period of oscillations is

$$T = 2\pi\sqrt{\frac{m}{k}} = 2 \times 3.14\sqrt{\frac{2}{196}} = 0.63 \text{ s}$$

The frequency of oscillations is

$$n = \frac{1}{T} = \frac{1}{0.63} = 1.59 \text{ Hz}$$

Exercise 19: A body of mass 2 kg attached to a spring makes S.H.M. with time-period 2.4 s. When the mass of the body is increased to 8 kg and set to make S.H.M., calculate the time-period of oscillations.

Solution: For the body of mass 2 kg, the time-period is

$$T = 2\pi\sqrt{\frac{m}{k}} \qquad \text{or} \qquad 2.4 = 2\pi\sqrt{\frac{2}{k}}$$

For the body of mass 8 kg, the time-period is

$$T' = 2\pi\sqrt{\frac{8}{k}}$$

From these equations, we have

$$\frac{T'}{2.4} = \frac{\sqrt{8/k}}{\sqrt{2/k}} = 2 \qquad \text{or} \qquad T' = 4.8 \text{ s}$$

5. Superposition of two parallel simple harmonic motions

Let us consider two simple harmonic motions having equal frequency ω and acting simultaneously on a particle parallel to each other as

$$y_1 = a_1 \sin(\omega t + \alpha_1) \qquad \text{and} \qquad y_2 = a_2 \sin(\omega t + \alpha_2)$$

Here, y_1 and y_2 are displacements, a_1 and a_2 the amplitudes, and α_1 and α_2 phase angles, respectively, of the motions. The two motions have equal frequency ω. Suppose these two vibrations act simultaneously on a body, the resultant displacement y is

$$y = y_1 + y_2 = a_1 \sin(\omega t + \alpha_1) + a_2 \sin(\omega t + \alpha_2)$$

$$= a_1(\sin \omega t \cos \alpha_1 + \cos \omega t \sin \alpha_1) + a_2(\sin \omega t \cos \alpha_2 + \cos \omega t \sin \alpha_2)$$

$$= \sin \omega t(a_1 \cos \alpha_1 + a_2 \cos \alpha_2) + \cos \omega t(a_1 \sin \alpha_1 + a_2 \sin \alpha_2) \quad (5.16)$$

Since a_1, a_2, α_1 and α_2 are constant, let us substitute as

$$A \cos \phi = a_1 \cos \alpha_1 + a_2 \cos \alpha_2 \qquad (5.17)$$

and

$$A \sin \phi = a_1 \sin \alpha_1 + a_2 \sin \alpha_2 \qquad (5.18)$$

On using equations (5.17) and (5.18) in (5.16), we have

$$y = A[\sin \omega t \, \cos \phi + \cos \omega t \, \sin \phi] = A \sin(\omega t + \phi)$$

This is equation of simple harmonic motion with amplitude A, frequency ω and phase angle ϕ. Hence, the resultant motion also has the same frequency as of the two motions. The amplitude A and phase angle ϕ can be determined in the following manner.

On squaring equations (5.17) and (5.18), and adding, we get

$$A^2 = (a_1 \cos \alpha_1 + a_2 \cos \alpha_2)^2 + (a_1 \sin \alpha_1 + a_2 \sin \alpha_2)^2$$

$$= a_1^2 \cos^2 \alpha_1 + a_2^2 \cos^2 \alpha_2 + 2a_1 a_2 \cos \alpha_1 \cos \alpha_2$$

$$+ a_1^2 \sin^2 \alpha_1 + a_2^2 \sin^2 \alpha_2 + 2a_1 a_2 \sin \alpha_1 \sin \alpha_2$$

or

$$A = [a_1^2 + a_2^2 + 2a_1 a_2 \cos(\alpha_1 - \alpha_2)]^{1/2} \tag{5.19}$$

On dividing equation (5.18) by (5.17), we get

$$\tan \phi = \frac{a_1 \sin \alpha_1 + a_2 \sin \alpha_2}{a_1 \cos \alpha_1 + a_2 \cos \alpha_2} \tag{5.20}$$

Equations (5.19) and (5.20) give values of A and ϕ.

Special cases:

(i) Suppose $(\alpha_1 - \alpha_2) = 0$, i.e, $\alpha_1 = \alpha_2 = \alpha$ (say). Then, we have

$$A = [a_1^2 + a_2^2 + 2a_1 a_2]^{1/2} = a_1 + a_2$$

and

$$\tan \phi = \frac{(a_1 + a_2) \sin \alpha}{(a_1 + a_2) \cos \alpha} = \tan \alpha \qquad \text{or} \qquad \phi = \alpha$$

(ii) Suppose $(\alpha_1 - \alpha_2) = \pi$, i.e, $\alpha_1 = \pi + \alpha_2$, and $a_1 \neq a_2$. Then, we have

$$A = [a_1^2 + a_2^2 - 2a_1 a_2]^{1/2} = |a_1 - a_2|$$

$$\tan \phi = \frac{a_1 \sin(\pi + \alpha_2) + a_2 \sin \alpha_2}{a_1 \cos(\pi + \alpha_2) + a_2 \cos \alpha_2)} = \frac{-a_1 \sin \alpha_2 + a_2 \sin \alpha_2}{-a_1 \cos \alpha_2 + a_2 \cos \alpha_2}$$

$$= \frac{(a_2 - a_1) \sin \alpha_2}{(a_2 - a_1) \cos \alpha_2} = \tan \alpha_2$$

Hence, $\phi = \alpha_2$.

(iii) Suppose $(\alpha_1 - \alpha_2) = \pi$, *i.e*, $\alpha_1 = \pi + \alpha_2$, and $a_1 = a_2$. Then the amplitude A of the resultant simple harmonic motion will be zero.

Exercise 20: Two simple harmonic motions acting simultaneously on a particle are

$$y_1 = 3\sin(\omega t + \pi/6) \qquad \text{and} \qquad y_2 = 4\sin(\omega t + \pi/3)$$

Calculate (i) amplitude, (ii) phase angle and (iii) time-period of the resultant motion.

Solution: For the simple harmonic motions

$$y_1 = 3\sin(\omega t + \pi/6) \qquad \text{and} \qquad y_2 = 4\sin(\omega t + \pi/3)$$

the frequencies are the same and therefore the time-periods are the same. Suppose, both the motions are acting simultaneously on a particle. Then, the resultant motion is

$$y = A\sin(\omega t + \phi)$$

(i) The amplitude A of the resultant motion is

$$A = [3^2 + 4^2 + 2(3)(4)\cos(\pi/6 - \pi/3)]^{1/2} = \sqrt{37}$$

(ii) The phase angle ϕ is

$$\tan\phi = \frac{3\sin\pi/6 + 4\sin\pi/3}{3\cos\pi/6 + 4\cos\pi/3} = \frac{3(0.5) + 4(\sqrt{3}/2)}{3(\sqrt{3}/2) + 4(0.5)} = 1.0796$$

Thus, we have

$$\phi = \tan^{-1}(1.0796) = 47.19°$$

(iii) The resultant time-period is the same as the individual time-periods.

Exercise 21: Two simple harmonic motions acting simultaneously on a particle are

$$y_1 = 2\sin(\omega t + \pi/2) \qquad \text{and} \qquad y_2 = 3\sin(\omega t + \pi/4)$$

Calculate (i) amplitude, (ii) phase angle and (iii) time-period of the resultant motion.

Solution: For the simple harmonic motions

$$y_1 = 2\sin(\omega t + \pi/2) \qquad \text{and} \qquad y_2 = 3\sin(\omega t + \pi/4)$$

the frequencies are the same and therefore the time-periods are the same. Suppose, both the motions are acting simultaneously on a particle. Then, the resultant motion is

$$y = A\sin(\omega t + \phi)$$

(i) The amplitude A of the resultant motion is

$$A = [2^2 + 3^2 + 2(2)(3)\cos(\pi/2 - \pi/4)]^{1/2} =$$

(ii) The phase angle ϕ is

$$\tan\phi = \frac{2\sin\pi/2 + 3\sin\pi/4}{2\cos\pi/2 + 3\cos\pi/4} = \frac{2 * 3(0.7071)}{3(0.7071)} = 1.9428$$

Thus, we have

$$\phi = \tan^{-1}(1.9428) = 62.76°$$

(iii) The resultant time-period is the same as the individual time-periods.

6. Superposition of two perpendicular simple harmonic motions (Lissajous figures)

Let us consider two simple harmonic motions having equal frequency ω and acting simultaneously on a particle perpendicular to each other as

$$x = a\sin(\omega t + \alpha) \tag{5.21}$$

and

$$y = b\sin\omega t \tag{5.22}$$

Here, x and y are displacements along two mutually perpendicular directions, say, along x and y axes, respectively. a and b, respectively, are the amplitudes of the motions. For convenience, the phase angle for one motion (along y-axis) is taken zero and therefore, α is the phase difference between the two motions. The two motions have equal frequency ω. From equation (5.22), we have

$$\sin\omega t = \frac{y}{b} \qquad \text{and therefore} \qquad \cos\omega t = \sqrt{1 - \frac{y^2}{b^2}} \tag{5.23}$$

From equation (5.21), we have

$$\frac{x}{a} = \sin\omega t \cos\alpha + \cos\omega t \sin\alpha \tag{5.24}$$

Using equation (5.23) in (5.24) and after rearranging, we get

$$\frac{x}{a} = \frac{y}{b}\cos\alpha + \sqrt{1 - \frac{y^2}{b^2}}\,\sin\alpha$$

or

$$\frac{x}{a} - \frac{y}{b}\cos\alpha = \sqrt{1 - \frac{y^2}{b^2}}\,\sin\alpha \qquad (5.25)$$

On squaring two sides of equation (5.26), we get

$$\frac{x^2}{a^2} + \frac{y^2}{b^2}\cos^2\alpha - \frac{2xy}{ab}\cos\alpha = \left(1 - \frac{y^2}{b^2}\right)\sin^2\alpha$$

or

$$\frac{x^2}{a^2} + \frac{y^2}{b^2} - \frac{2xy}{ab}\cos\alpha = \sin^2\alpha \qquad (5.26)$$

This is general equation of an ellipse. The resultant motion of the particle will however depend on the value of α. The paths of the motion obtained for various values of α are known as the Lissajous figures.

(i) When $\alpha = 0$ or 2π, we have $\cos\alpha = 1$, $\sin\alpha = 0$ and the equation (5.26) becomes

$$\frac{x^2}{a^2} + \frac{y^2}{b^2} - \frac{2xy}{ab} = 0 \qquad \text{or} \qquad \frac{x}{a} - \frac{y}{b} = 0 \qquad \text{or} \qquad y = \frac{b}{a}x$$

This represents the equation of a straight line. That is, the particle moves simple harmonically along a straight line AB, shown in the following figure.

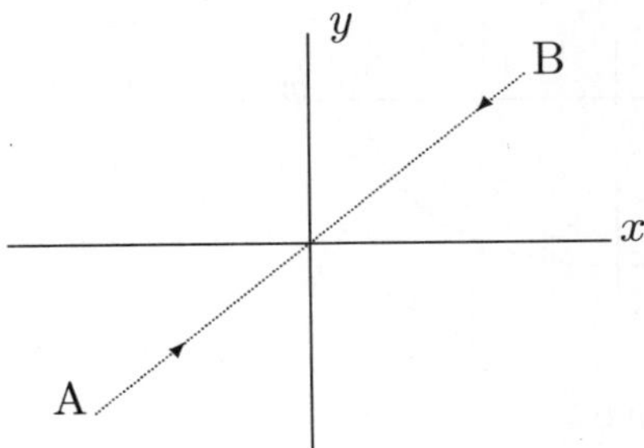

(ii) When $\alpha = \pi$, we have $\cos\alpha = -1$, $\sin\alpha = 0$ and the equation (5.26) becomes

$$\frac{x^2}{a^2} + \frac{y^2}{b^2} + \frac{2xy}{ab} = 0 \qquad \text{or} \qquad \frac{x}{a} + \frac{y}{b} = 0 \qquad \text{or} \qquad y = -\frac{b}{a}\, x$$

This represents the equation of a straight line. That is, the particle moves simple harmonically along a straight line CD, shown in the following figure.

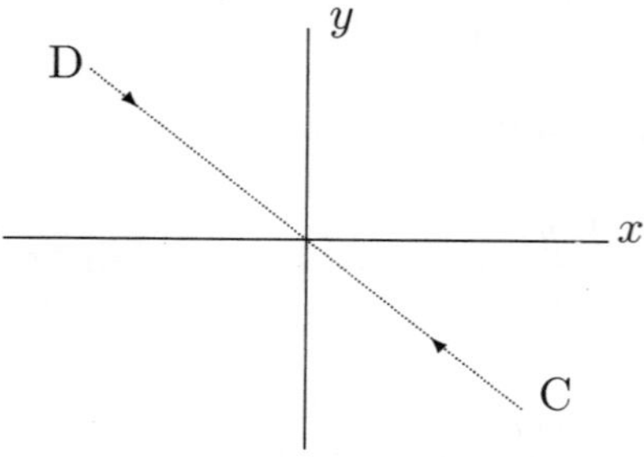

(iii) When $\alpha = \pi/2$ or $3\pi/2$, we have $\cos\alpha = 0$, $\sin\alpha = 1$ and the equation (5.26) becomes

$$\frac{x^2}{a^2} + \frac{y^2}{b^2} = 1$$

This represents the equation of an ellipse. That is, the particle moves simple harmonically along an ellipse, shown in Figure

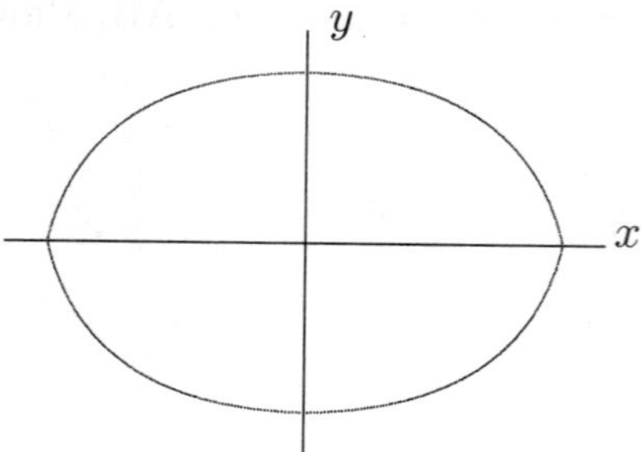

Further, when $a = b$, we have

$$x^2 + y^2 = a^2$$

This represents the equation of a circle. That is, the particle moves simple harmonically along a circle, shown in the following figure.

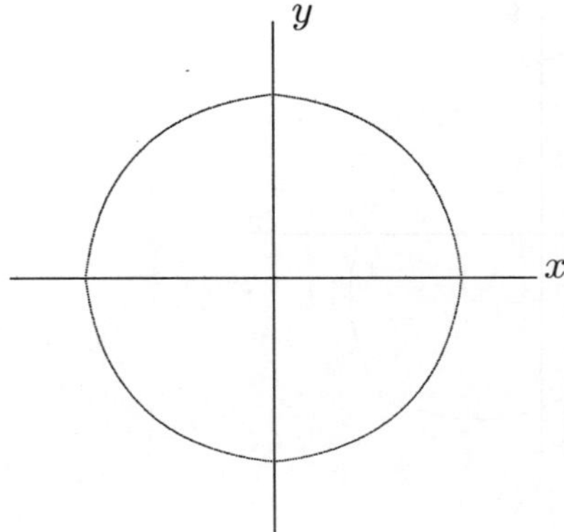

(iv) When $\alpha = \pi/4$ or $7\pi/4$, we have $\cos\alpha = 1/\sqrt{2}$, $\sin\alpha = 1/\sqrt{2}$ and the equation (5.26) becomes

$$\frac{x^2}{a^2} + \frac{y^2}{b^2} - \frac{2xy}{ab}\frac{1}{\sqrt{2}} = \frac{1}{2}$$

This represents the equation of an ellipse, shown in the following figure.

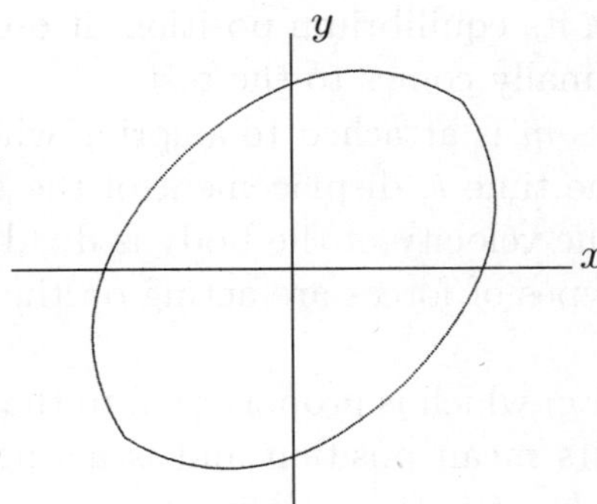

(v) When $\alpha = 3\pi/4$ or $5\pi/4$, we have $\cos\alpha = -1/\sqrt{2}$, $\sin\alpha = 1/\sqrt{2}$ and the equation (5.26) becomes

$$\frac{x^2}{a^2} + \frac{y^2}{b^2} + \frac{2xy}{ab}\frac{1}{\sqrt{2}} = \frac{1}{2}$$

This represents the equation of an ellipse, shown in the following figure.

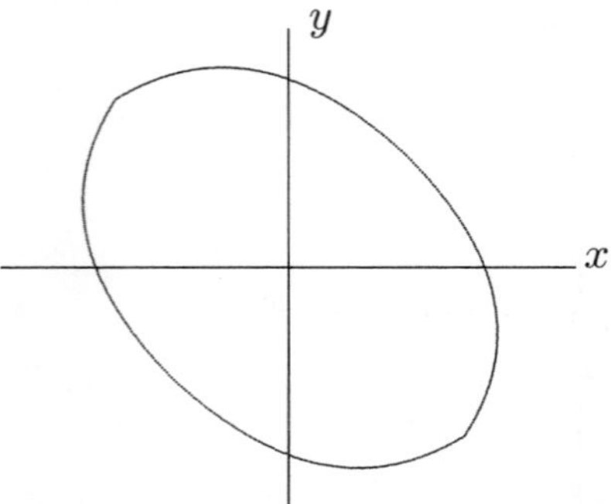

7. Damped harmonic oscillator

In the preceding sections, we discussed about free oscillations where no frictional force or resistance is offered to the oscillating body. Consequently, the body keeps on oscillating indefinitely and such oscillations are known as the free oscillations. But, in the real situations, there is always some resistance offered to the oscillating body. In real sense when a body is set into oscillations, its amplitude continuously decreases due to fractional resistance working on the body and therefore the vibrations will die after some finite time. Such motion is said to be damped by the friction and is known as the damped oscillations. Thus, when a pendulum is displaced from its equilibrium position, it oscillates with a decreasing amplitude and finally comes to the rest.

    Consider a body of mass m is attached to a spring whose force constant is k. Suppose at some time t, displacement of the body from its equilibrium position is x, the velocity of the body is dx/dt and acceleration d^2x/dt^2. Here, two types of forces are acting on the body:

(i) The restoring force $(-kx)$ which is proportional to the displacement x of the body from its mean position and is acting towards the mean position. Here, k is the force constant.

(ii) The damping force $(-q\,dx/dt)$ which is proportional to the velocity of the body and is directed opposite to the direction of the motion of the body. Here, k is coefficient of damping.

Hence, the total force F acting on the body is

$$F = -kx - q\,\frac{dx}{dt} \tag{5.27}$$

With acceleration $\mathrm{d}^2x/\mathrm{d}t^2$ of the body having mass m, the force is

$$F = m\,\frac{\mathrm{d}^2x}{\mathrm{d}t^2} \tag{5.28}$$

From equations (5.27) and (5.28), we have

$$m\,\frac{\mathrm{d}^2x}{\mathrm{d}t^2} = -kx - q\,\frac{\mathrm{d}x}{\mathrm{d}t}$$

or

$$\frac{\mathrm{d}^2x}{\mathrm{d}t^2} + 2s\,\frac{\mathrm{d}x}{\mathrm{d}t} + \omega^2 x = 0 \tag{5.29}$$

where $s = q/2m$ is the damping factor and $\omega = \sqrt{k/m}$ the angular frequency. This is differential equation of second order. Let us assume its solution to be of the form

$$x = A\,\mathrm{e}^{\alpha t} \tag{5.30}$$

so that

$$\frac{\mathrm{d}x}{\mathrm{d}t} = \alpha A\,\mathrm{e}^{\alpha t} \qquad \text{and} \qquad \frac{\mathrm{d}^2x}{\mathrm{d}t^2} = \alpha^2 A\,\mathrm{e}^{\alpha t} \tag{5.31}$$

Using equations (5.30) and (5.31) in (5.29), we have

$$\alpha^2 A\,\mathrm{e}^{\alpha t} + 2s\alpha A\,\mathrm{e}^{\alpha t} + \omega^2 A\,\mathrm{e}^{\alpha t} = 0$$

or

$$A\,\mathrm{e}^{\alpha t}(\alpha^2 + 2s\alpha + \omega^2) = 0$$

Since $A\,\mathrm{e}^{\alpha t} \neq 0$, we have

$$\alpha^2 + 2s\alpha + \omega^2 = 0$$

This is quadratic equation in α and its roots are

$$\alpha_1 = -s + \sqrt{s^2 - \omega^2} \qquad \text{and} \qquad \alpha_2 = -s - \sqrt{s^2 - \omega^2}$$

Therefore, the general solution of equation (5.29) is

$$x = A_1\,\mathrm{e}^{(-s+\sqrt{s^2-\omega^2})t} + A_2\,\mathrm{e}^{(-s-\sqrt{s^2-\omega^2})t} \tag{5.32}$$

where A_1 and A_2 are arbitrary constants. The actual solution depends on the situation whether (i) $s^2 > \omega^2$, (ii) $s^2 = \omega^2$ or (iii) $s^2 < \omega^2$.

(i) Case A: $s^2 > \omega^2$. Here, $\sqrt{(s^2 - \omega^2)}$ is real and less than s. Therefore, both the exponents $(-s + \sqrt{s^2 - \omega^2})$ and $(-s - \sqrt{s^2 - \omega^2})$ are negative. Further, $(-s - \sqrt{s^2 - \omega^2})$ is more negative as compared to $(-s + \sqrt{s^2 - \omega^2})$. Consequently, the factor $(-s - \sqrt{s^2 - \omega^2})$ will dominate and the displacement x decreases exponentially with time to zero without performing any oscillations. Such motion is said to be over-damped and may be expressed as

$$x = A_2 \, e^{(-s - \sqrt{s^2 - \omega^2})t}$$

(ii) Case B: $s^2 = \omega^2$. Here, we have

$$x = (A_1 + A_2) \, e^{-st}$$

In this case also the displacement x decreases exponentially with time to zero without performing any oscillations. Such motion is known as the critical damped, as it is the limit of the damped motion without oscillation.

(iii) Case C: $s^2 < \omega^2$. Here, $\sqrt{s^2 - \omega^2}$ is imaginary and can be written as

$$\sqrt{s^2 - \omega^2} = i\beta$$

where $\beta = \sqrt{\omega^2 - s^2}$ and $i = \sqrt{-1}$. Thus, equation (5.32) becomes

$$x = e^{-st}[A_1 \, e^{i\beta t} + A_2 \, e^{-i\beta t}]$$

$$= e^{-st}[A_1(\cos \beta t + i \sin \beta t) + A_2(\cos \beta t - i \sin \beta t)]$$

$$= e^{-st}[(A_1 + A_2) \cos \beta t + i(A_1 - A_2) \sin \beta t] \tag{5.33}$$

Suppose $(A_1 + A_2) = A \sin \delta$ and $i(A_1 - A_2) = A \cos \delta$. Here, A and δ are constants. Then equation (5.33) becomes

$$x = e^{-st}[A \sin \delta \, \cos \beta t + A \cos \delta \, \sin \beta t] = A \, e^{-st} \cos(\beta t + \delta)$$

Substituting the value of β here, we have

$$x = A \, e^{-st} \cos[\sqrt{(\omega^2 - s^2)} \, t + \delta)] \tag{5.34}$$

This is equation of oscillatory motion with amplitude $A \, e^{-st}$. Obviously, the amplitude decreases with time t exponentially. However, the decay of amplitude depends on the damping factor s. Such motion is known as under damped. Examples of such motion are the motion of a pendulum in air, the motion of the coil of the ballistic galvanometer.

7.1 Energy-decay

By definition, the energy of an oscillator is proportional to the square of the amplitude of motion. In the damped oscillator, the amplitude decays exponentially with time as e^{-st}. Hence, the energy of the oscillator decays as e^{-2st}. The decay of energy depends on the damping factor s. The decay is fast for large value of s.

7.2 Logarithmic decrement

The rate at which the amplitude in the damped motion decays is measured through logarithmic decrement. The amplitude of damped harmonic motion is $A\,e^{-st}$. Therefore, at $t = 0$, the amplitude is the maximum $A = A_0$. Suppose A_1, A_2, A_3, ... be the amplitudes at the time $t = T$, $2T$, $3T$, ..., respectively, where T is the time-period of oscillations. Thus, we have

$$A_1 = A\,e^{-sT} \qquad\qquad A_2 = A\,e^{-2sT} \qquad\qquad A_3 = A\,e^{-3sT} \qquad\qquad \ldots$$

This gives

$$\frac{A_0}{A_1} = \frac{A_1}{A_2} = \frac{A_2}{A_3} = \ldots = e^{sT} = e^{\lambda}$$

Here, λ is known as the logarithmic decrement. By taking natural logarithm, we have

$$\ln\frac{A_0}{A_1} = \ln\frac{A_1}{A_2} = \ln\frac{A_2}{A_3} = \ldots = \lambda$$

Thus, the logarithmic decrement is the natural logarithm of the ratio between two successive amplitudes which are separated by one time-period.

7.3 Relaxation time

It is the time taken by a damped harmonic oscillator for decaying total mechanical energy by a factor $1/e$ of its initial value. The mechanical energy of the oscillator with amplitude $A\,e^{-st}$ is

$$E = \frac{1}{2}\,m\omega^2 (A\,e^{-st})^2$$

At $t = 0$, we have

$$E_0 = \frac{1}{2}\,m\omega^2 A^2$$

Thus, we have

$$E = E_0 \, e^{-2st} \tag{5.35}$$

Suppose, τ be the relaxation time, then at $t = \tau$, we have

$$E = \frac{E_0}{e}$$

It gives

$$\frac{E_0}{e} = E_0 \, e^{-2s\tau} \qquad \text{or} \qquad e = e^{2s\tau}$$

Therefore, we have

$$2s\tau = 1 \qquad \text{or} \qquad \tau = \frac{1}{2s}$$

Thus, equation (5.35) becomes

$$E = E_0 \, e^{-t/\tau}$$

This equation shows how the energy decays exponentially with time t.

Exercise 22: The relaxation time for a damped harmonic oscillator is 25 s. Calculate the time required to decrease the amplitude of the oscillator to $1/e$ times of initial value.

Solution: With initial amplitude A_0, the amplitude of damped harmonic oscillator is

$$A = A_0 \, e^{-st}$$

For the relaxation time τ, we have

$$\tau = \frac{1}{2s} \qquad \text{or} \qquad s = \frac{1}{2\tau} = \frac{1}{2 \times 25} = 0.02 \ \text{s}^{-1}$$

Suppose after a time t', we have $A = A_0/e$. Then,

$$\frac{A_0}{e} = A_0 \, e^{-st'} \qquad \text{or} \qquad e = e^{st'}$$

Thus,

$$st' = 1 \qquad \text{or} \qquad t' = \frac{1}{s} = \frac{1}{0.02} = 50 \ \text{s}$$

Exercise 23: A damped oscillator starting from rest, has the first amplitude of 50 cm which reduces to 5 cm in that direction after 100 oscillations. When the period of each oscillation is 3.2 s, calculate the damping constant.

Solution: Suppose the time-period of oscillation is T. From the mean position, the time taken to reach the first amplitude is $T/4$. Thus, we have

$$A_1 = A_0\, e^{-sT/4} \qquad \text{and} \qquad A_2 = A_0\, e^{-s(T/4+T)}$$

and therefore,

$$A_{n+1} = A_0\, e^{-s(T/4+nT)}$$

Thus, we have

$$\frac{A_1}{A_{n+1}} = e^{snT}$$

Using the values, we have

$$\frac{50}{5} = e^{100s(3.2)} \qquad \text{or} \qquad \ln 10 = 320\ s$$

Thus, we have damping constant $s = 0.0072$.

8. Forced oscillations

In the preceding section, we have discussed oscillations where a body oscillates with its own frequency in absence of any external force. A different situation arises when an oscillating body is placed in an external periodic force. Such oscillations are known as the forced oscillations. In this case, the body finally oscillates under the influence of the external force and not with its natural frequency. So the forced oscillations can also be defined as the oscillations in which the body oscillates with frequency other than the natural frequency. This other frequency is due to some external periodic force. The tuning fork is one of the examples of the forced oscillations.

Suppose a body of mass m is connected to a mass-less, inextensible and perfectly flexible string. When it is displaced from its mean (equilibrium) position and released, it starts to oscillate to and fro about the mean position. Suppose at some time t, displacement of the body from its mean position is x, the velocity of the body is dx/dt and acceleration d^2x/dt^2. Here, the body experiences three kinds of forces:

(i) The restoring force $(-kx)$ which is proportional to the displacement x of the body from its mean position and is acting towards the mean position. Here, k is the force constant.

(ii) The damping force $(-q\,dx/dt)$ which is proportional to the velocity of the body and is directed opposite to the direction of the motion of the body. Here, k is coefficient of damping.

(iii) The external periodic force $F_0\sin pt$ with angular frequency p.

Hence, the total force F acting on the body is

$$F = -kx - q\,\frac{dx}{dt} + F_0\sin pt \tag{5.36}$$

With acceleration d^2x/dt^2 of the body, the force is

$$F = m\,\frac{d^2x}{dt^2} \tag{5.37}$$

From equations (5.36) and (5.37), we have

$$m\,\frac{d^2x}{dt^2} = -kx - q\,\frac{dx}{dt} + F_0\sin pt$$

or

$$\frac{d^2x}{dt^2} + 2s\,\frac{dx}{dt} + \omega^2 x = f\sin pt \tag{5.38}$$

where $s = q/2m$ is the damping factor, $\omega = \sqrt{k/m}$ the angular frequency and $f = F_0/m$. Because of damping force, the body will loose its energy. The external force will however give energy periodically to the body, and finally, the body will oscillate with the frequency of the external force. Equation (5.38) is differential equation of second order. Since the body finally oscillates with the frequency of the applied external force. Let us assume that in the steady state, the solution of equation (5.38) be of the form

$$x = A\sin(pt - \delta) \tag{5.39}$$

where A the amplitude of oscillations and δ is known as the phase angle. We have

$$\frac{dx}{dt} = Ap\cos(pt - \delta) \qquad \text{and} \qquad \frac{d^2x}{dt^2} = -Ap^2\sin(pt - \delta) \tag{5.40}$$

Using equations (5.39) and (5.40) in (5.38), we have

$$-Ap^2\sin(pt - \delta) + 2sAp\cos(pt - \delta) + \omega^2 A\sin(pt - \delta)$$

$$= f\sin[(pt - \delta) + \delta]$$

or

$$A\sin(pt-\delta)[\omega^2-p^2]+2sAp\cos(pt-\delta)=f\sin(pt-\delta)\ \cos\delta$$

$$+f\cos(pt-\delta)\ \sin\delta \tag{5.41}$$

Comparing the coefficients of $\sin(pt-\delta)$ and $\cos(pt-\delta)$ on the two sides of this equation, we get

$$A(\omega^2-p^2)=f\cos\delta \tag{5.42}$$

and

$$2sAp=f\sin\delta \tag{5.43}$$

Squaring equations (5.42) and (5.43) and adding them we get

$$A^2(\omega^2-p^2)^2+4s^2A^2p^2=f^2$$

or

$$A=\frac{f}{\sqrt{(\omega^2-p^2)^2+4s^2p^2}} \tag{5.44}$$

Dividing equation (5.43) by (5.42), we get

$$\tan\delta=\frac{2sp}{(\omega^2-p^2)}$$

or the phase angle

$$\delta=\tan^{-1}\left[\frac{2sp}{(\omega^2-p^2)}\right] \tag{5.45}$$

Equations (5.44) and (5.45) show that both the amplitude and the phase angle of forced oscillations depend on (ω^2-p^2). That is, they depend on both the driving frequency p and the natural frequency ω of the oscillator. The amplitude and phase angle for various situations are discussed in the following.

(i) Case A: Suppose the driving frequency p is very small as compared to the natural frequency ω, *i.e.*, $p<<\omega$. Then the amplitude of oscillations is

$$A=\frac{f}{\sqrt{(\omega^2-p^2)^2+4s^2p^2}}\approx\frac{f}{\omega^2}$$

Using here expressions for f and ω, we have

$$A = \frac{F_0/m}{k/m} = \frac{F_0}{k}$$

It shows that the amplitude of the forced oscillations depends on the force constant k of the spring and the magnitude F_0 of the applied external force. The phase angle is

$$\delta = \tan^{-1}\left[\frac{2sp}{(\omega^2 - p^2)}\right] = \tan^{-1}\left[\frac{2sp}{\omega^2}\right]$$

Since $\omega^2 >> p^2$, we have $2sp/\omega^2 \to 0$ and $\delta \to 0$. Thus, for this case, the driving force and the displacement are in phase.

(ii) **Case B:** Suppose the driving frequency p is equal to the natural frequency ω, $i.e.$, $p = \omega$. Such frequency is known as the resonant frequency. Then the amplitude of oscillations is

$$A = \frac{f}{\sqrt{(\omega^2 - p^2)^2 + 4s^2p^2}} = \frac{f}{2sp}$$

Substituting the expressions for f and s, we have

$$A = \frac{F_0/m}{(q/m)p} = \frac{F_0}{qp}$$

It shows that the amplitude A of forced oscillations depends on the damping coefficient q and the frequency p of applied force. The phase angle is

$$\delta = \tan^{-1}\left[\frac{2sp}{(\omega^2 - p^2)}\right] = \tan^{-1}[\infty] = \frac{\pi}{2}$$

Thus, the displacement lags behind the applied force by a phase angle of $\pi/2$ radians.

(iii) **Case C:** Suppose the driving frequency is very large as compared to the natural frequency, $i.e.$, $p >> \omega$. Then the amplitude of oscillations is

$$A = \frac{f}{\sqrt{(\omega^2 - p^2)^2 + 4s^2p^2}} = \frac{f}{\sqrt{p^4 + 4s^2p^2}} = \frac{f}{p^2}$$

Substituting the expression for f, we have

$$A = \frac{F_0}{mp^2}$$

It shows that the amplitude of the forced oscillations depends on the amplitude F_0 and frequency p of the applied force. The phase angle is

$$\delta = \tan^{-1}\left[\frac{2sp}{(\omega^2 - p^2)}\right] = \tan^{-1}\left[\frac{2sp}{-p^2}\right]$$

$$= \tan^{-1}\left[\frac{2s}{-p}\right] = \tan^{-1}[-0] = \pi$$

For this case, the displacement lags behind the force by a phase angle of π radians.

8.1 Resonance

In order to understand the phenomenon of resonance, we consider example of the forced oscillations which occur at the frequency of the external force with driving frequency p and not the natural frequency ω of the vibrating system. We have seen that the amplitude of oscillations depends on the relationship between p and ω. The amplitude is

$$A = \frac{f}{\sqrt{(\omega^2 - p^2)^2 + 4s^2 p^2}}$$

For $p = 0$, we have

$$A_0 = \frac{f}{\omega^2}$$

and for $p = \omega$, we have

$$A_\omega = \frac{f}{2s\omega}$$

Therefore,

$$\frac{A_\omega}{A_0} = \frac{\omega}{2s}$$

It shows that ω must be larger than $2s$. For the given value of ω, when the damping factor s decreases, the amplitude A increases and the increase is the largest near $p = \omega$. When there is no damping ($s = 0$), the amplitude becomes infinite at $p = \omega$. In Figure 4, we have shown variation of amplitude of forced oscillations as a function of p/ω for three values (small, medium and large) of damping factor s. Figure shows that the amplitude decreases when p is away from ω. The amplitude is the maximum around $p = \omega$. The situation $p = \omega$ is known as the resonance. Further, when the damping is small, the amplitude of the forced oscillations increases rapidly as p approaches to ω. The amplitude reaches its maximum when $p = \omega$. For medium damping also, the amplitude increases but it does not increase so rapidly near $p = \omega$. As the damping increases, the resonance frequency is displaced slightly away from the natural frequency.

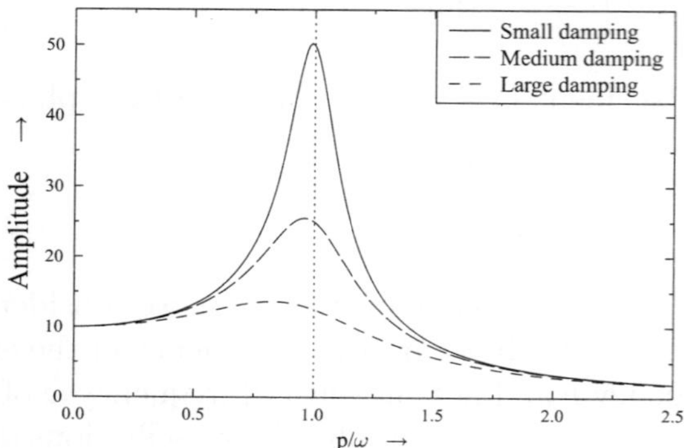

Figure 4: Variation of amplitude versus p/ω for three values of damping factor s.

8.2 Quality factor

Suppose the energy E of an oscillator varies with its frequency p in such a manner that with the increase of frequency, the energy first increases and then decreases, as shown in Figure 5. Suppose the energy is maximum E_{max} at the frequency $p = p_0$. Thus, the energy decreases on both sides of the frequency p_0. If at the frequencies p_1 $(p_1 < p_0)$ and and p_2 $(p_2 > p_0)$, the energy is $E_{max}/2$ (*i.e.*, half of the maximum). The band-width is expressed as

$$\Delta p = |p_1 - p_2|$$

and the quality factor Q is expressed as

$$Q = \frac{p_0}{\Delta p} = \frac{p_0}{|p_1 - p_2|}$$

Suppose a particle of mass m is oscillating with frequency p and amplitude A. The average energy is expressed as

$$E = \frac{1}{2}\, mp^2 A^2$$

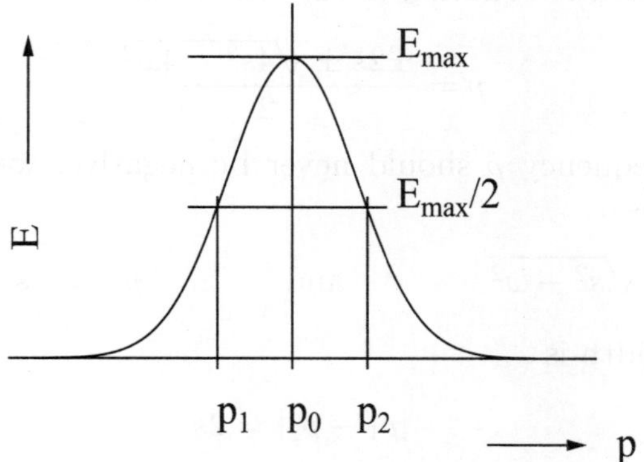

Figure 5: Variation of energy E with frequency p.

Using the value of the amplitude A, we have

$$E = \frac{1}{2} \frac{mp^2 f^2}{[(\omega^2 - p^2)^2 + 4s^2 p^2]}$$

The energy will be maximum $E = E_{max}$ when $p = p_0$. Thus, we have

$$E_{max} = \frac{1}{2} \frac{mp_0^2 f^2}{[(\omega^2 - p_0^2)^2 + 4s^2 p_0^2]}$$

That is, when $(\omega^2 - p_0^2) = 0$, *i.e.*,

$$p_0 = \omega = \sqrt{\frac{k}{m}}$$

we have

$$E_{max} = \frac{1}{2} \frac{mp_0^2 f^2}{4s^2 p_0^2} = \frac{mf^2}{8s^2}$$

Now, when we have $(\omega^2 - p^2)^2 = 4s^2 p^2$, the energy is

$$E_1 = \frac{1}{2} \frac{mp^2 f^2}{8s^2 p^2} = \frac{mf^2}{16s^2} = \frac{E_{max}}{2}$$

It shows that the energy is half of the maximum, when we have

$$(\omega^2 - p^2)^2 = 4s^2 p^2 \qquad \text{or} \qquad \omega^2 - p^2 = \pm 2sp$$

Thus, we have

$$p^2 \pm 2sp - \omega^2 = 0$$

For this quadratic equation having the roots

$$p = \frac{\pm 2s \pm \sqrt{4s^2 + 4\omega^2}}{2}$$

Since the frequency p should never be negative, we have two possible values of p as

$$p_1 = s + \sqrt{s^2 + \omega^2} \qquad \text{and} \qquad p_2 = -s + \sqrt{s^2 + \omega^2}$$

The band width is

$$|p_1 - p_2| = 2s$$

Here, p_0 is the frequency when the energy is the maximum, and p_1 and p_2 the frequencies when the energy is half of the maximum. Thus, the quality factor Q is

$$Q = \frac{p_0}{|p_1 - p_2|} = \frac{\sqrt{k/m}}{2s} = \frac{\sqrt{k/m}}{q/m} = \frac{\sqrt{km}}{q}$$

9. Multiple choice questions

1. For a particle making simple harmonic motion, the velocity is the maximum at the

   (i) equilibrium position   (ii) extreme position

   (iii) mid-way of the equilibrium position and extreme

   (iv) 1/3 of amplitude from the equilibrium position

   Ans. (i)

2. For a particle making simple harmonic motion, the phase difference between displacement and velocity is

   (i) 0   (ii) $\pi/2$ (iii) π (iv) 2π

   Ans. (ii)

3. For a particle making simple harmonic motion, the total energy is the same as at

   (i) equilibrium position   (ii) extreme position

   (iii) mid-way of the equilibrium position and extreme

   (iv) 1/3 of amplitude from the equilibrium position

   Ans.

4. For a particle making simple harmonic motion, the potential energy is the maximum at

    (i) equilibrium position   (ii) extreme position

    (iii) mid-way of the equilibrium position and extreme

    (iv) 1/3 of amplitude from the equilibrium position

    Ans. (ii)

5. For a particle making simple harmonic motion, the kinetic energy is the maximum at

    (i) equilibrium position   (ii) extreme position

    (iii) mid-way of the equilibrium position and extreme

    (iv) 1/3 of amplitude from the equilibrium position

    Ans. (i)

6. For a particle making simple harmonic motion with amplitude A, the kinetic energy is equal to the potential energy at

    (i) $A/2$ (ii) A (iii) $A/\sqrt{2}$ (iv) $A/\sqrt{3}$

    Ans.

7. For a particle making simple harmonic motion, if the amplitude is doubled, the new time-period would be

    (i) same   (ii) double   (iii) half   (iv) four times

    Ans.

8. In which of the following oscillations, the amplitude varies with time.

    (i) damped oscillator   (ii) forced oscillator

    (iii) simple harmonic oscillator   (iv) none of them

    Ans.

9. The unit of spring constant in SI system is

    (i) N m^2 (ii) N m   (iii) N/m   (iv) N/m^2

    Ans. (iii)

10. Suppose the energy E of an oscillator varies with its frequency p in such a manner that the energy increases with frequency and then decreases. If the energy is the maximum E_{max} at frequency $p = p_0$, the energy decreases on both sides of the frequency p_0. If

at the frequencies p_1 and p_2, the energy is $E_{max}/2$. The quality factor Q is

(i) $\dfrac{p_0}{|p_1 + p_2|}$ (ii) $\dfrac{p_0}{|p_1 - p_2|}$ (iii) $\dfrac{|p_1 + p_2|}{p_0}$ (iv) $\dfrac{|p_1 - p_2|}{p_0}$

Ans. (ii)

10. Problems and questions

1. Describe the simple harmonic motion of a particle.

2. A particle of mass 200 g is executing S.H.M. with time-period of π seconds. The displacement of the particle from its equilibrium position is 80 cm at time $\pi/8$ second. Calculate total energy of the particle.

3. A particle executes S.H.M. of time-period 20 s and amplitude 10 cm. Calculate maximum velocity of the particle.

4. A particle of mass m executes S.H.M. under a restoring force with force constant k. Obtain expressions for displacement, velocity and acceleration of the particle.

5. Suppose displacement x of a particle of mass m executing S.H.M. is

$$x = A\sin(\omega t + \delta)$$

Derive expressions for kinetic energy and potential energy of the particle. Show that total energy of the particle is conserved.

6. Show that during the S.H.M., the kinetic energy of the particle changes into the potential energy of the particle and vice-versa.

7. Derive expression for the time-period of simple pendulum.

8. Show that the time period of a simple pendulum does not depend on its mass.

9. Describe the horizontal motion of a mass attached to a perfectly elastic and mass-less spring.

10. Describe the vertical motion of a mass attached to a perfectly elastic and mass-less spring.

11. Suppose two simple harmonic motions

$$x_1 = a\sin(\omega t + \alpha) \qquad \text{and} \qquad x_2 = b\sin\omega t$$

    are acting simultaneously on a particle parallel to each other. Describe the resultant motion of the particle.

12. Suppose two simple harmonic motions

$$x = a\sin(\omega t + \alpha) \qquad \text{and} \qquad y = b\sin\omega t$$

    are acting simultaneously on a particle perpendicular to each other. Describe the resultant motion of the particle.

13. Describe motion of a damped harmonic oscillator.

14. Describe motion of an oscillator under an external force.

15. Show that for the forced oscillations, when the frequency of the external force is much smaller than the natural frequency of oscillations show that displacement and the driving force are in phase.

16. Write notes on the following

    (i) Harmonic motion

    (ii) Time-period

    (iii) Frequency

    (iv) Transverse and longitudinal modes of a wave

    (v) Simple pendulum

    (vi) Lissajous figures

    (vii) Resonance

    (viii) Quality factor

VI. Sound

Sound is produced through the vibrations of a body. For example, when a gong of a bell is struck with a hammer, sound is produced. The bell is set into vibration and the sound propagates through air. These vibrations reach the ear and the ear drum is set into vibrations.

A tuning fork can be set into vibration by striking one of its prongs against a rubber pad. The vibrations of the prongs of the tuning fork can be seen.

It has been proved with the help of experiments that for propagation of a sound wave, medium is necessarily required. That is, in absence of a medium, no sound wave can travel. The mode of propagation of a sound wave is longitudinal.

1. Wave motion

Wave motion is a form of disturbance which propagates through a medium due to repeated periodic motion of the particles of the medium about their mean positions, the disturbance being handed over from one particle to the next. For example, when a stone is dropped into a pond containing water, waves are produced at the point where the stone strikes the water in the pond. The waves travel outward and the particles of water vibrate only up and down about their mean positions. The water particles do not travel along with the wave. Similarly, when a tuning fork is set into vibration, it produces waves in air. These waves travel from one place to other, but the particles of air vibrate about their mean positions.

Let us try to understand - what is propagated in a wave motion? The answer to this question is that, the physical condition due to a disturbance produced at some point in the medium propagates to other point in the medium. In each wave, the particles in the medium vibrate about their mean positions. Thus, in case of wave motion, it is not matter that propagates, but it is only state of motion of the matter that is propagated. It is a form of dynamic condition that is propagated from one place to other place in the medium.

According to the laws of physics, any dynamic condition is related to momentum and energy. It may therefore be said that in a wave motion, momentum and energy are transferred from one place to other place. However, no matter is transferred from any place to other place.

1.1 Modes of propagation

The waves have two modes of propagation: (i) transverse motion and (ii) longitudinal motion. In the transverse motion, the particles of the medium vibrate in the direction perpendicular to the direction of propagation of the wave. In the longitudinal motion, the particles of the medium vibrate in the direction parallel to the direction of propagation of the wave.

2. Equation of simple harmonic wave

During propagation of a simple harmonic wave, let the displacement of a particle A at any instant of time t is expressed as

$$y = A \sin \omega t$$

Consider another particle B at a distance x from the particle A. Here, we assume that the wave propagates from A towards B with a velocity v. The displacement at B is

$$y = A \sin(\omega t - \alpha) \tag{6.1}$$

where α denotes the phase difference between the particles A and B. A phase difference 2π corresponds to the path difference λ (wavelength of the wave). Thus, we have

$$\frac{\alpha}{x} = \frac{2\pi}{\lambda} \qquad \text{or} \qquad \alpha = \frac{2\pi x}{\lambda}$$

Now, for a time-period T, the angular frequency ω is

$$\omega = \frac{2\pi}{T} = \frac{2\pi v}{\lambda}$$

Here, we have used the relations $v = \nu\lambda$ and $\nu = 1/T$. Using the expressions for α and ω in equation (6.1), we have

$$y = A \sin \left(\frac{2\pi v t}{\lambda} - \frac{2\pi x}{\lambda} \right) \qquad \text{or} \qquad y = A \sin \frac{2\pi}{\lambda}(vt - x)$$

This equation represents the simple harmonic wave propagating in the positive direction, *i.e.*, from left towards right. It gives displacement at a distance x at time t. Similarly for a particle at a distance x in the negative direction, the displacement is expressed as

$$y = A \sin \frac{2\pi}{\lambda}(vt + x)$$

Notice that there is change in the sign of distance x.

2.1 Differential form of equation of motion

Consider the equation of simple harmonic motion of a wave moving in the positive direction

$$y = A \sin \frac{2\pi}{\lambda}(vt - x) \tag{6.2}$$

On differentiating equation (6.2) with respect to time t, we have

$$\frac{dy}{dt} = \frac{2\pi A v}{\lambda} \cos \frac{2\pi}{\lambda}(vt - x) \tag{6.3}$$

Again differentiating equation (6.3) with respect to t, we have

$$\frac{d^2 y}{dt^2} = -\frac{4\pi^2 A v^2}{\lambda^2} \sin \frac{2\pi}{\lambda}(vt - x) \tag{6.4}$$

On differentiating equation (6.2) with respect to displacement x, we have

$$\frac{dy}{dx} = -\frac{2\pi A}{\lambda} \cos \frac{2\pi}{\lambda}(vt - x) \tag{6.5}$$

Again differentiating equation (6.5) with respect to x, we have

$$\frac{d^2 y}{dx^2} = -\frac{4\pi^2 A}{\lambda^2} \sin \frac{2\pi}{\lambda}(vt - x) \tag{6.6}$$

From equations (6.3) and (6.5), we have

$$\frac{dy}{dt} = -v \, \frac{dy}{dx} \tag{6.7}$$

From equations (6.4) and (6.6), we have

$$\frac{d^2 y}{dt^2} = v^2 \, \frac{d^2 y}{dx^2} \tag{6.8}$$

Equation (6.8) represents the differential form of wave motion.

2.2 Particle velocity and wave velocity

Consider the equation of simple harmonic motion of a wave moving in the positive direction

$$y = A \sin \frac{2\pi}{\lambda}(vt - x) \tag{6.9}$$

Here, v is the velocity of the wave. The velocity u of a particle vibrating about its mean position is

$$u = \frac{dy}{dt} = \frac{2\pi A v}{\lambda} \cos \frac{2\pi}{\lambda}(vt - x) \tag{6.10}$$

The velocity of the particle depends on time. The maximum velocity is

$$u_{max} = \frac{2\pi A v}{\lambda}$$

Thus, we have

$$\text{maximum velocity of particle} = \frac{2\pi A}{\lambda} \text{ velocity of wave}$$

The acceleration f of a particle is

$$f = \frac{d^2 y}{dt^2} = -\frac{4\pi^2 A v^2}{\lambda^2} \sin \frac{2\pi}{\lambda}(vt - x) \qquad \text{or} \qquad f = -\frac{4\pi^2 A v^2}{\lambda^2} y$$

The acceleration of the particle also depends on time. The maximum acceleration is

$$f_{max} = -\frac{4\pi^2 A v^2}{\lambda^2} A$$

The negative sign shows that the acceleration of the particle is always directed towards its mean position. On differentiation equation (6.9) with respect to x, we have

$$\frac{dy}{dx} = -\frac{2\pi A}{\lambda} \cos \frac{2\pi}{\lambda}(vt - x) \tag{6.11}$$

From equations (6.10) and (6.11), we get

$$\frac{dy}{dt} = -v \frac{dy}{dx} \qquad \text{or} \qquad u = -v \frac{dy}{dx}$$

This is the relation between the particle-velocity and wave-velocity.

Exercise 1: For a simple harmonic wave propagating through a medium, the displacement y of a particle (in cm) at any instant t is expressed as

$$y = 20 \sin \frac{2\pi}{50}(32000\, t - 20)$$

Calculate the amplitude of the vibrating particle, wave velocity, wavelength, frequency and time-period.

Solution: General equation of a simple harmonic wave is

$$y = A \sin \frac{2\pi}{\lambda}(vt - x)$$

On comparison of this equation with the given equation, we have

Amplitude $A = 20$ cm

Wavelength $\lambda = 50$ cm

Wave velocity $v = 32000$ cm/s

Frequency $n = v/\lambda = 32000/50 = 640$ Hz

Time-period $T = 1/n = 1/640$ sec

Exercise 2: A simple harmonic wave having amplitude 8 cm is propagating in the positive x-direction. At any instant of time, for a particle at a distance 10 cm from the origin, the displacement is 6 cm and for a particle at a distance 25 cm from the origin, the displacement is 4 cm. calculate the wavelength of the wave.

Solution: General equation of a simple harmonic wave is

$$y = A \sin \frac{2\pi}{\lambda}(vt - x) \qquad \text{or} \qquad \frac{y}{A} = \sin 2\pi \left(\frac{t}{T} - \frac{x}{\lambda}\right)$$

In the first case, we have

$$\frac{6}{8} = \sin 2\pi \left(\frac{t}{T} - \frac{10}{\lambda}\right) \tag{6.12}$$

In the second case, we have

$$\frac{4}{8} = \sin 2\pi \left(\frac{t}{T} - \frac{25}{\lambda}\right) \tag{6.13}$$

From equation (6.12), we get

$$2\pi \left(\frac{t}{T} - \frac{10}{\lambda}\right) = \sin^{-1}(0.75) = \frac{48.59\pi}{180} \tag{6.14}$$

From equation (6.13), we get

$$2\pi \left(\frac{t}{T} - \frac{25}{\lambda}\right) = \sin^{-1}(0.5) = 30° = \frac{\pi}{6} \text{ rad} \tag{6.15}$$

Subtracting equation (6.15) from (6.14), we get

$$\frac{50\pi}{\lambda} - \frac{20\pi}{\lambda} = \frac{48.59\pi}{180} - \frac{30\pi}{180}$$

Therefore, we have

$$\lambda = \frac{180 \times 30}{18.59} = 290.5 \text{ cm}$$

Exercise 3: A simple harmonic wave is propagating with velocity of 30 cm/s. At a time $t = 0$, the displacement of a particle is

$$y = 4 \, \sin 2\pi \left(\frac{x}{100}\right)$$

Find the equation for the displacement of the particle at the time $t = 2$ s.

Solution: General equation of a simple harmonic wave is

$$y = A \, \sin \frac{2\pi}{\lambda}(vt - x) \qquad \text{or} \qquad y = A \, \sin 2\pi \left(\frac{t}{T} - \frac{x}{\lambda}\right)$$

At $t = 0$, we have

$$y = -A \, \sin 2\pi \left(\frac{x}{\lambda}\right)$$

On comparison of this equation with the given equation, we get $a = -4$ and $\lambda = 100$ cm. We are given, $v = 30$ cm/s. For these values, the equation of the wave can be expressed as

$$y = A \, \sin \frac{2\pi}{\lambda}(vt - x) = -4 \, \sin \frac{2\pi}{100}(30 \, t - x)$$

At $t = 2$, we have

$$y = -4 \, \sin \frac{2\pi}{100}(30 \times 2 - x) = 4 \, \sin \frac{2\pi}{100}(x - 60)$$

This is the required equation.

Exercise 4: A simple harmonic wave is expressed as

$$y = 10 \, \sin \left(\frac{2\pi t}{T} + \alpha\right)$$

The time-period for the wave is 30 s. At time $t = 0$, the displacement is 5 cm. Calculate (i) phase angle at $t = 7.5$ s, and (ii) the phase different between two positions at a time interval of 6 s.

Solution: General equation of a simple harmonic wave is

$$y = A \sin \left(\frac{2\pi t}{T} + \alpha \right)$$

Given $a = 10$ cm and $T = 30$ s, and we have

$$y = 10 \sin \left(\frac{2\pi t}{30} + \alpha \right)$$

At $t = 0$, we have $y = 5$ cm. Therefore, we have

$$5 = 10 \sin \alpha \qquad \text{or} \qquad \sin \alpha = 0.5$$

Therefore, $\alpha = \pi/6$. The phase angle is

$$\phi = \frac{2\pi t}{30} + \alpha = \frac{2\pi t}{30} + \frac{\pi}{6}$$

(i) At $t = 7.5$ s, the phase angle is

$$\phi = \frac{2\pi 7.5}{30} + \frac{\pi}{6} = \frac{2\pi}{3}$$

(ii) The phase angles ϕ_1 and ϕ_2 at the times t_1 and t_2, respectively, are

$$\phi_1 = \frac{2\pi t_1}{30} + \frac{\pi}{6} \qquad \text{and} \qquad \phi_2 = \frac{2\pi t_2}{30} + \frac{\pi}{6}$$

The phase difference is

$$\phi_2 - \phi_1 = \frac{2\pi (t_2 - t_1)}{30}$$

For $(t_2 - t_1) = 6$ s, the phase difference is

$$\phi_2 - \phi_1 = \frac{2\pi \times 6}{30} = \frac{2\pi}{5}$$

3. Velocity of sound

In order to obtain the velocity of sound in a medium, let us consider a long cylindrical tube of area of cross-section a which is filled with the material of the medium, as shown in Figure 1. Let a sound wave propagates through the medium from left towards right. Because of the propagation of the wave, there will be condensations and rarefactions in the medium. At the place of condensation, the density will increase and the volume will decrease whereas at the place of rarefaction, the density will decrease and the volume will increase. Let us consider two cross sections A and B of the tube. For a small region, the pressure, velocity and density of the medium will remain constant. The values of the pressure, velocity and density of the medium will be different at different cross-sections of the tube.

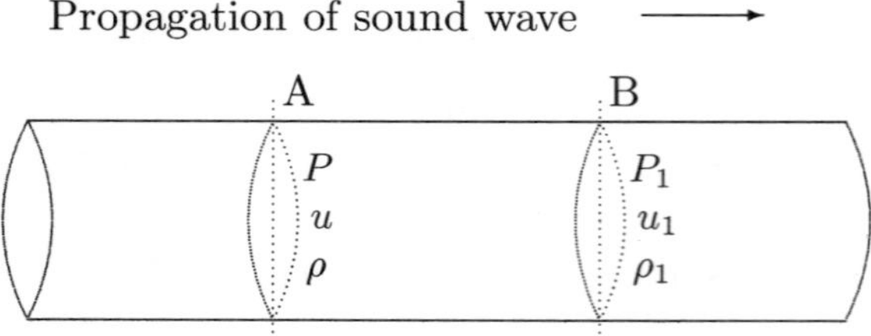

Figure 1: A sound wave propagates through the medium filled in a cylindrical tube of area of cross-section a.

Let the cross-section A corresponds to the normal region of the medium with the pressure P, density ρ, and velocity of the wave u and the cross-section B corresponds to the condensation or rarefaction of the medium with the pressure P_1, density ρ_1, and velocity of the wave u_1.

(i) The mass of medium entering the cross-section A in one second is

$$= au\rho$$

(ii) The mass of medium leaving the cross-section B in one second is

$$= au_1\rho_1$$

As the average density of the medium does not change with time, the mass m entering at the cross-section A will be equal to that leaving at the cross-section B. Thus, we have

$$m = au\rho = au_1\rho_1 \qquad \text{or} \qquad u\rho = u_1\rho_1$$

(i) When B is the region of condensation, its density will be higher than that in the normal region, and we have

$$\rho_1 > \rho \qquad \text{and} \qquad u_1 < u$$

(ii) When B is the region of rarefaction, its density will be lower than that in the normal region, and we have

$$\rho_1 < \rho \qquad \text{and} \qquad u_1 > u$$

The momentum entering at A per second is mu

The momentum leaving at B per second is mu_1

The change in momentum per second is $mu - mu_1$

This change in momentum per second is due to the pressure difference $(P_1 - P)$ at the two cross-sections. For the area of cross-section a, therefore, the force applied is

$$F = (P_1 - P)a$$

Following the Newton's second law of motion, the rate of change of momentum is equal to the applied force. Thus, we have

$$(P_1 - P)a = mu - mu_1 \qquad \text{or} \qquad (P_1 - P)a = mu\left(1 - \frac{u_1}{u}\right)$$

On using $m = au\rho$ here, we get

$$(P_1 - P) = u^2\rho\left(1 - \frac{u_1}{u}\right)$$

On using $u_1\rho_1 = u\rho$, *i.e.*, $u_1 = u\rho/\rho_1$ here, we get

$$(P_1 - P) = u^2\rho\left(1 - \frac{u\rho/\rho_1}{u}\right) \qquad \text{or} \qquad u^2\rho = \frac{\rho_1(P_1 - P)}{(\rho_1 - \rho)} \qquad (6.16)$$

The bulk modulus of elasticity E of the medium is expressed as

$$E = \frac{\text{stress}}{\text{volume strain}} = \frac{P_1 - P}{(V - V_1)/V}$$

where V and V_1 are volumes at the cross-sections A and B, respectively. Now, the density is inversely proportional to the volume

$$V \propto \frac{1}{\rho} \qquad \text{and} \qquad V_1 \propto \frac{1}{\rho_1}$$

Therefore, we have

$$\frac{V - V_1}{V} = \frac{\rho_1 - \rho}{\rho_1}$$

Thus, the bulk modulus of elasticity is

$$E = \frac{\rho_1(P_1 - P)}{(\rho_1 - \rho)} \qquad (6.17)$$

From equations (6.16) and (6.17), we have

$$u^2\rho = E \qquad \text{or} \qquad u = \sqrt{\frac{E}{\rho}}$$

This expression for the velocity of sound is true only in the case of plane waves where the original disturbance has very small amplitude. This condition is satisfied by the sound waves of ordinary intensity. This relation also holds good for the velocity of simple harmonic plane waves propagating through homogeneous isotropic mediums, *viz,* solids, liquids and gases.

3.1　Newton's formula for velocity of sound

As we have seen that the velocity of sound u in a medium (solid, liquid or gas) depends on the elasticity E and density ρ of the medium as

$$u = \sqrt{\frac{E}{\rho}}$$

Newton found that when the medium is air, the elasticity E can be replaced by the pressure P of the air. He assumed when the sound waves travel through the air, the temperature of air remains constant. Further, he assumed that the air can be considered as an ideal gas. For the isothermal condition in an ideal gas, the Boyle's law can be applied. At the point of condensation, the pressure increases and the volume decreases, whereas at the point of rarefaction, the pressure decreases and the volume increases. Let the initial pressure and volume of the air are P and V, respectively. Suppose, during condensation, the increase in pressure is p and the decrease in volume is v. Thus, the final pressure and volume of the air are $(P + p)$ and $(V - v)$, respectively. From the Boyle's law. we have

$$PV = (P + p)(V - v) \qquad \text{or} \qquad PV = PV + pV - Pv - pv$$

On neglecting very small term pv, we get

$$pV = Pv \qquad \text{or} \qquad P = \frac{pV}{v} \tag{6.18}$$

From the definition of the elasticity of the medium, we have

$$E = \frac{\text{change in pressure}}{\text{change in volume/original volume}} = \frac{p}{v/V} = \frac{pV}{v} \tag{6.19}$$

From equations (6.18) and (6.19), we have

$$E = P$$

Therefore, the velocity of sound propagating through air having pressure P and density ρ is

$$u = \sqrt{\frac{P}{\rho}}$$

Taking normal pressure $P = 76$ cm of Hg, we have

$$P = h\rho g = 76 \times 13.6 \times 981 = 1.014 \times 10^6 \text{ dyne/cm}^2$$

The density of air at the N.T.P. is $\rho = 1.293 \times 10^{-3}$ gm/cm^3. Hence, the velocity of sound is

$$u = \sqrt{\frac{P}{\rho}} = \sqrt{\frac{1.014 \times 10^6}{1.293 \times 10^{-3}}} = 28004 \text{ cm/s}$$

The velocity of sound determined from various experiments at the N.T.P. is 33200 cm/s. Thus, there is a difference of about 5196 cm/s between the theoretical and experimental values. So large difference cannot be attributed to experimental errors. Newton could not provide any explanation for the difference. The explanation was later on provided by Laplace.

3.1.1 Laplace correction

According to Laplace, when a sound wave travels through a medium, there are condensations and rarefactions in the medium. At the time of condensation, the particles come close to each other and are heated up. At the time of rarefaction, the particles go apart from each other and the temperature cools down. Consequently, the temperature in the medium does not remain constant when a sound wave travels through air or any gas. Hence, the process is not isothermal, but is adiabatic. The total quantity of heat of the system as a whole remains constant. During the adiabatic process, we have

$$PV^\gamma = \text{constant}$$

where γ is the ratio of two specific heats of the gas. For the initial pressure P and volume V, and final pressure $(P+p)$ and volume $(V-v)$ at the condensation (as discussed above), we have

$$PV^\gamma = (P+p)(V-v)^\gamma \qquad \text{or} \qquad PV^\gamma = P\left(1 + \frac{p}{P}\right)V^\gamma\left(1 - \frac{v}{V}\right)^\gamma$$

Using the Binomial expansion, we have

$$1 = \left(1 + \frac{p}{P}\right)\left(1 - \frac{\gamma v}{V}\right) \qquad \text{or} \qquad 1 = 1 + \frac{p}{P} - \frac{\gamma v}{V} - \frac{\gamma v p}{VP}$$

On neglecting the term $\gamma v p / V P$, we get

$$\frac{p}{P} = \frac{\gamma v}{V} \qquad \text{or} \qquad \gamma P = \frac{pV}{v}$$

Thus, the elasticity of the medium is

$$E = \frac{pV}{v} = \gamma P$$

Therefore, the velocity of sound propagating through air having pressure P and density ρ is

$$u = \sqrt{\frac{\gamma P}{\rho}} \qquad\qquad (6.20)$$

The value of γ for the air is 1.42. Therefore, the velocity of sound is

$$u = \sqrt{\frac{\gamma P}{\rho}} = \sqrt{\frac{1.42 \times 1.014 \times 10^6}{1.293 \times 10^{-3}}} = 33370.6 \text{ cm/s}$$

This value of the velocity of sound in air is in good agreement with the experimental value. Therefore, the correct expression for the velocity of sound through a gas is expressed by equation (6.20).

3.1.2   Effect of temperature

The density of air or gas changes with temperature. If ρ_0 is the density of the air at 0 C, then the velocity u_0 at 0 C is

$$u_0 = \sqrt{\frac{\gamma P}{\rho_0}}$$

If ρ_t is the density of the air at t C, then the velocity u_t at at t C is

$$u_t = \sqrt{\frac{\gamma P}{\rho_t}}$$

Therefore, we have

$$\frac{u_t}{u_0} = \sqrt{\frac{\rho_0}{\rho_t}}$$

For an ideal gas with pressure P, volume V, temperature T, we have

$$PV = RT \qquad\qquad \text{or} \qquad\qquad \frac{Pm}{\rho} = RT$$

where m is mass of the gas and ρ the density. For two temperatures 0 C and t C (corresponding temperatures T_0 and T in Kelvin), we have

$$\frac{Pm}{\rho_0} = RT_0 \qquad \text{and} \qquad \frac{Pm}{\rho_t} = RT$$

Therefore, we have

$$\frac{\rho_0}{\rho_t} = \frac{T}{T_0}$$

Thus, we have

$$\frac{u_t}{u_0} = \sqrt{\frac{T}{T_0}}$$

It shows that the velocity of sound in air or gas is directly proportional to the square root of the absolute temperature of the medium.

3.1.3 Effect of pressure

For an ideal gas with pressure P, volume V, temperature T, we have

$$PV = RT \qquad \text{or} \qquad \frac{P}{\rho} = \frac{RT}{m}$$

where m is mass of the gas and ρ the density. It shows that for a given mass m of the gas and given temperature T, the ratio of pressure and density is constant. That is,

$$\frac{P}{\rho} = \text{constant}$$

It shows that the velocity of sound in the air does not depend on the pressure in the gas, as with the change of pressure there is change in the density in proportion.

3.1.4 Effect of density

We have seen that when the mass of the gas is given, the ratio of pressure and density is constant. When the mass of the gas is allowed to vary such that its pressure P is constant, at two densities ρ_1 and ρ_2, the respective velocities u_1 and u_2 are given as

$$u_1 = \sqrt{\frac{\gamma P}{\rho_1}} \qquad \text{and} \qquad u_2 = \sqrt{\frac{\gamma P}{\rho_2}}$$

Therefore, we have

$$\frac{u_1}{u_2} = \sqrt{\frac{\rho_2}{\rho_1}}$$

It shows that the velocity of sound in air or gas is inversely proportional to the square root of the density.

Exercise 5: The velocity of sound in air at 17 C is 330 m/s. What will be the velocity of sound in the air at 157 C when the pressure is doubled?

Solution: The velocity of sound does not depend on the pressure in the medium. At two temperatures t_1 C and t_2 C, the ratio of respective velocities u_1 and u_2 is

$$\frac{u_2}{u_1} = \sqrt{\frac{273 + t_2}{273 + t_1}}$$

Using the values, we have

$$\frac{u_2}{330} = \sqrt{\frac{273 + 157}{273 + 17}}$$

On solving, we get $u_2 = 401.8$ m/s.

4. Stationary waves

Let us consider two simple harmonic waves A and B of equal frequency, equal time-period and equal magnitude, propagating in the opposite directions along a straight line. The wave A is propagating in the right direction whereas the wave B is propagating in the left direction. The resultant of such waves is known as the stationary wave or standing wave. The formation of stationary waves is due to the superposition of the two waves propagating in the opposite directions in the medium. The formation of the stationary waves can be represented graphically in the following manner. Here, we shall look at the particles 1, 2, 3, 4, 5, 6, 7, 8, 9 of the medium, situated on a straight line parallel the the directions of motions. Let T be the time-period of the waves.

(i) Suppose at the time $t = 0$, the two waves A and B are as shown in Figure 2. The resultant displacement C of the waves A and B is a straight line. All the particles of the medium are on the line.

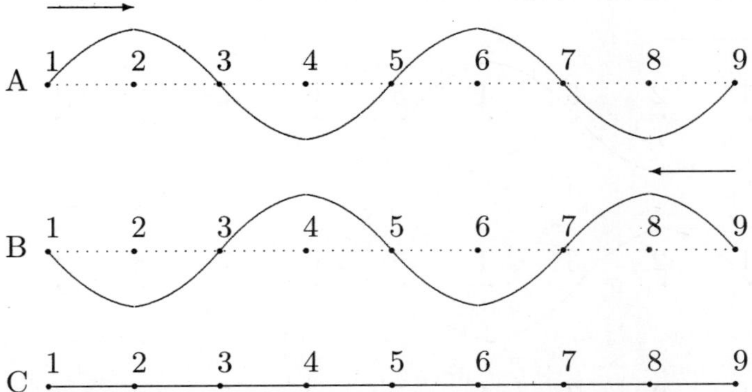

Figure 2: Two waves A and B propagating in opposite directions and their resultant C at the time $t = 0$.

(ii) At the time $t = T/4$, the wave A will advance through a distance $\lambda/4$ in the right direction whereas the wave B will advance through a distance $\lambda/4$ in the left direction, as shown in Figure 3. The resultant displacement C of the waves A and B is no more a straight line. However, the particles 2, 4, 6, 8 are still at their positions whereas the particles 1, 3, 5, 7 are at their extreme displacements.

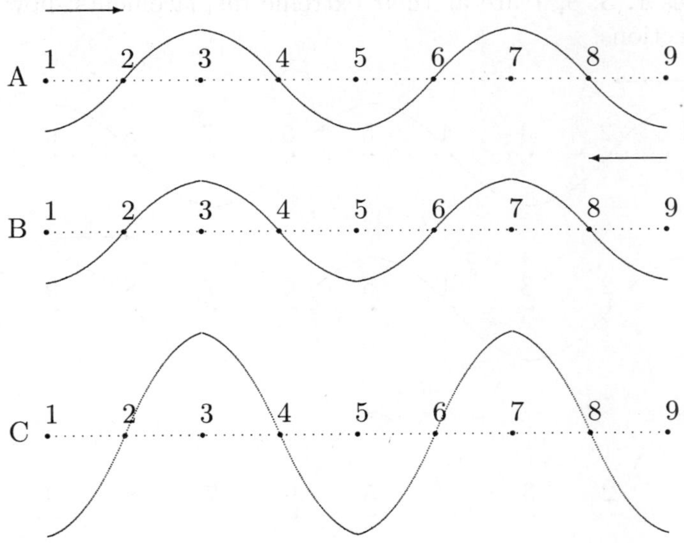

Figure 3: Two waves A and B propagating in opposite directions and their resultant C at the time $t = T/4$.

(iii) At the time $t = T/2$, the wave A will advance through a distance $\lambda/2$ in the right direction whereas the wave B will advance through a distance $\lambda/2$ in the left direction, as shown in Figure 4. The resultant displacement C of the A and B waves is again a straight line.

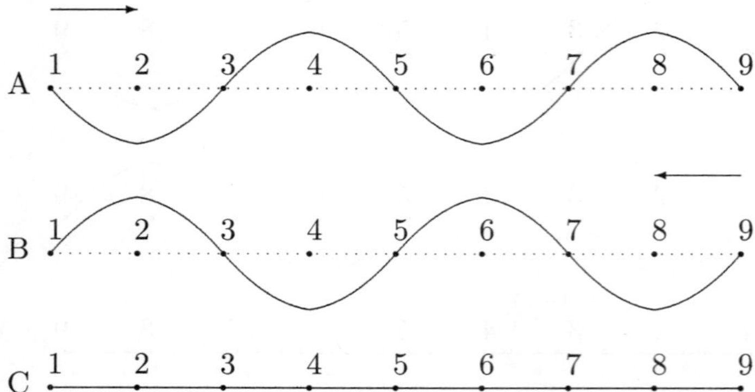

Figure 4: Two waves A and B propagating in opposite directions and their resultant C at the time $t = T/2$.

(iv) At the time $t = 3T/4$, the wave A will advance through a distance $3\lambda/4$ in the right direction whereas the wave B will advance through a distance $3\lambda/4$ in the left direction, as shown in Figure 5. The resultant displacement C of the A and B waves is no more a straight line. However, the particles 2, 4, 6, 8 are still at their positions whereas the particles 1, 3, 5, 7 are at their extreme displacements; now in the reverse directions.

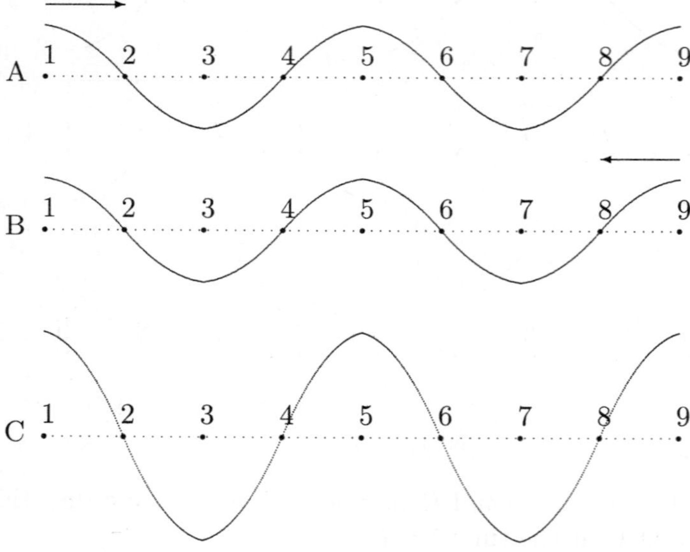

Figure 5: Two waves A and B propagating in opposite directions and their resultant C at the time $t = 3T/4$.

(v) At the time $t = T$, the wave A will advance through a distance λ in the right direction wheres the wave B will advance through a distance λ in the left direction, and the situation will be the same as shown in Figure 2. That is, the phenomenon now repeats.

From the patterns discussed above, it is clear that some particles of the medium, such as 2, 4, 6, 8, etc. always remain fix at their positions. On the other side, the particles 1, 3, 5, 7, etc. oscillate periodically about their mean positions on the two sides with double the magnitude of each wave. It appears as the wave pattern is stationary in the space. The resultant displacement patterns of the particles A at different times is as shown in Figure 6

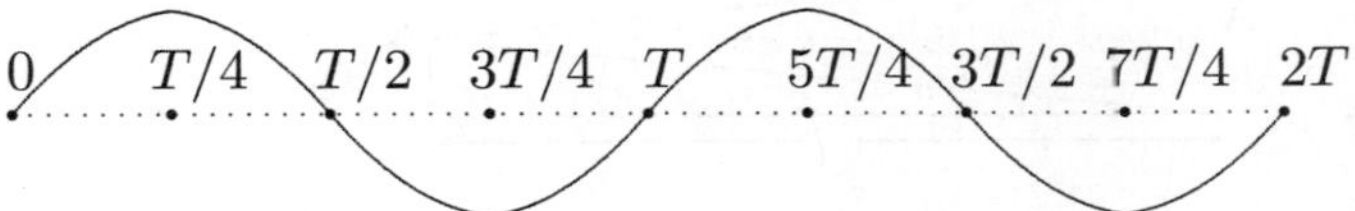

Figure 6: Displacement of particle 1 versus time.

The positions of the particles 2, 4, 6, 8 etc. which always remain at the fixed position are known as the nodes. Node is the position of zero displacement and maximum strain.

The positions of the particles 1, 3, 5, 7 etc. which oscillate in a simple harmonic motion with the maximum amplitude (twice the amplitude of each wave) are known as the antinodes. At the antinodes, the strain is the minimum. The distance between two consecutive nodes or between two antinodes is $\lambda/2$.

5. Standing waves in air column

We have various kinds of wind instruments which have a column of air, called the resonator. In a wind instrument, stationary longitudinal waves are produced. The vibrations are produced by the mouth piece and the mouth piece is different for different instruments. The mouth piece acts as a source and supplies the required energy to maintain the vibrations in an air-column. Depending on the nature of excitations, the organ pipes may be of two types:

(i) organ pipe closed at one end
(ii) organ pipe opened at both ends

The nature of reflected wave is different in the open-end or closed-end organ pipes. In case of a closed-end organ pipe, the reflection takes place at a rigid wall whereas in case of an open-end organ pipe, the reflection takes place at a yielding wall. In case of a closed-end organ pipe, the closed-end has a node whereas in case of an open-end organ pipe, the open-ends have anti-nodes.

5.1   Organ pipe closed at one end

In this kind of organ pipe, one end of the pipe is open and the other is close. Suppose the length of organ pipe is l. For the fundamental frequency of vibration of the air column, we have as shown in the following figure.

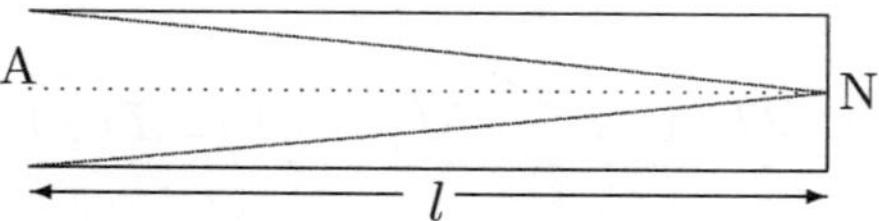

Therefore, we have

$$l = \frac{\lambda_1}{4} \qquad \text{or} \qquad \lambda_1 = 4l$$

where λ_1 is wavelength of the wave. For the velocity v of the sound wave, the frequency is

$$n_1 = \frac{v}{\lambda_1} = \frac{v}{4l}$$

For the first overtone, we have as in the following figure.

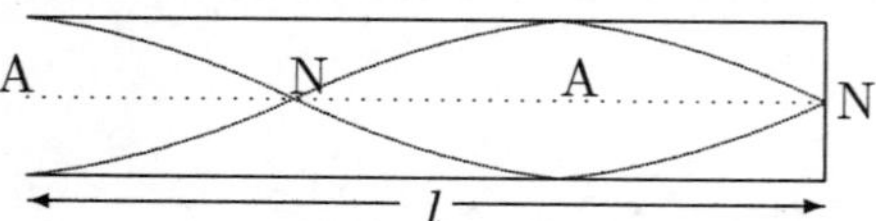

Therefore, we have

$$l = \frac{3\lambda_2}{4} \qquad \text{or} \qquad \lambda_2 = 4l/3$$

The corresponding frequency is

$$n_2 = \frac{v}{\lambda_2} = \frac{3v}{4l} = 3n_1$$

For the second overtone, we have as in the following figure.

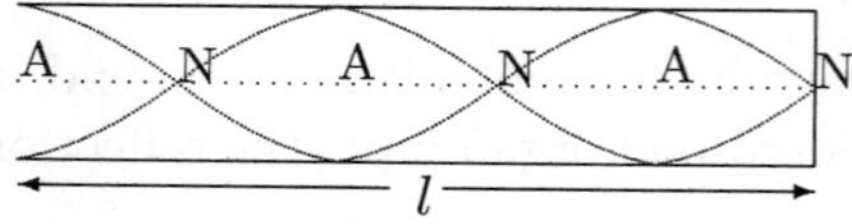

Therefore, we have

$$l = \frac{5\lambda_3}{4} \qquad \text{or} \qquad \lambda_3 = 4l/5$$

The corresponding frequency is

$$n_3 = \frac{v}{\lambda_3} = \frac{5v}{4l} = 5n_1$$

Thus, in case of a closed-end organ pipe, only the odd harmonics are formed and the even harmonics are absent. The fundamental frequency is $v/4l$.

5.2 Organ pipe opened at both ends

In this kind of organ pipe, both the ends of the pipe are open. In an open end organ pipe, anti-nodes are formed at both the ends. Suppose the length of organ pipe is l. For the fundamental frequency of vibration of the air column, we have as in the following figure.

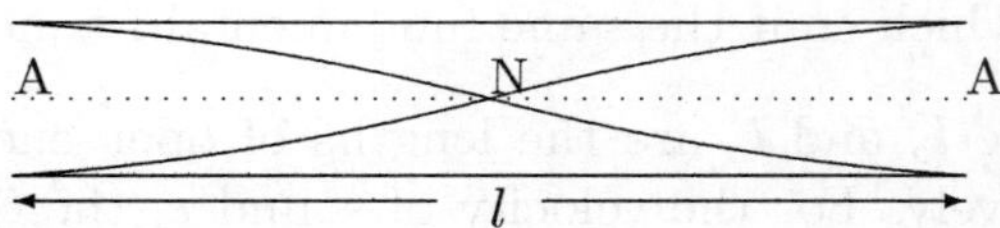

Therefore, we have

$$l = \frac{\lambda_1}{2} \qquad \text{or} \qquad \lambda_1 = 2l$$

where λ_1 is wavelength of the wave. For the velocity v of the sound wave, the frequency is

$$n_1 = \frac{v}{\lambda_1} = \frac{v}{2l}$$

For the first overtone, we have as in the following figure.

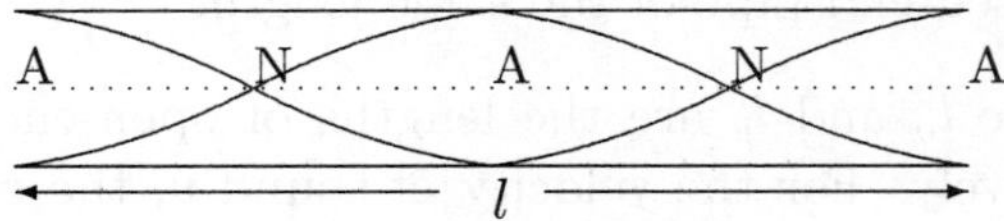

Therefore, we have

$$l = \lambda_2$$

The corresponding frequency is

$$n_2 = \frac{v}{\lambda_2} = \frac{v}{l} = 2n_1$$

For the second overtone, we have as in the following figure.

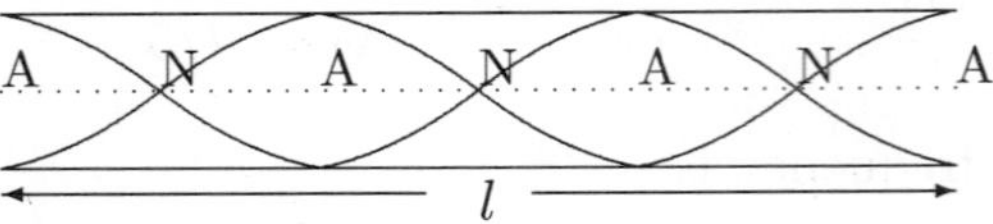

Therefore, we have

$$l = \frac{3\lambda_3}{2} \qquad \text{or} \qquad \lambda_3 = 2l/3$$

The corresponding frequency is

$$n_3 = \frac{v}{\lambda_3} = \frac{3v}{2l} = 3n_1$$

Thus, in case of an open-end organ pipe, we have the frequencies n_1, $2n_1$, $3n_1$, etc. Thus, all the harmonics are present.

Exercise 6: Calculate the ratio of the lengths of an open-end pipe and a closed-end pipe which emit the same fundamental frequency.

Solution: Suppose l_o and l_c are the lengths of open-end and closed-end pipes, respectively. For the velocity of sound v, the corresponding fundamental frequencies n_o and n_c are

$$n_o = \frac{v}{2l_o} \qquad \text{and} \qquad n_c = \frac{v}{4l_c}$$

For $n_o = n_c$, we have

$$\frac{v}{2l_o} = \frac{v}{4l_c} \qquad \text{and} \qquad \frac{l_o}{l_c} = 2$$

Exercise 7: Calculate the ratio of fundamental frequencies of an open-end pipe and a closed-end pipe of the same length.

Solution: Suppose l_o and l_c are the lengths of open-end and closed-end pipes, respectively. For the velocity of sound v, the corresponding fundamental frequencies n_o and n_c are

$$n_o = \frac{v}{2l_o} \qquad \text{and} \qquad n_c = \frac{v}{4l_c}$$

Therefore, we have

$$\frac{n_o}{n_c} = \frac{v/2l_o}{v/4l_c} = \frac{2l_c}{l_o}$$

For $l_o = l_c$, we have

$$\frac{n_o}{n_c} = 2$$

Exercise 8: Two open organ pipes of lengths 55 cm and 55.5 cm produce 3 beats/sec. Calculate the velocity of sound in air.

Solution: Here, $l_1 = 55$ cm and $l_2 - 55.5$ cm. For the velocity v of sound in air, the fundamental frequencies are

$$n_1 = \frac{v}{4l_1} = \frac{v}{110} \qquad \text{and} \qquad n_2 = \frac{v}{4l_2} = \frac{v}{111}$$

We have $n_1 - n_2 = 3$. Therefore,

$$3 = \frac{v}{110} - \frac{v}{111}$$

On solving, we get $v = 36630$ cm/s.

Exercise 9: Two tuning forks A and B produce 6 beats/sec. The tuning fork A resounds with a closed air column of 10 cm length whereas the tuning fork B resounds with an open column of 25 cm length. Calculate the velocity of sound in air.

Solution: Suppose v is the velocity of sound in air. The frequency of tuning fork A is

$$n_1 = \frac{v}{4l_1} = \frac{v}{4(10)}$$

The frequency of tuning fork B is

$$n_2 = \frac{v}{2l_2} = \frac{v}{2(25)}$$

Thus, we have

$$n_1 - n_2 = \frac{v}{40} - \frac{v}{50}$$

We are given $n_1 - n_2 = 6$. Thus, we have

$$6 = \frac{v}{40} - \frac{v}{50}$$

On solving this relation, we get $v = 1200$ cm/s.

Exercise 10: An open organ pipe filled with air has a fundamental frequency of 460 Hz. The fundamental frequency of another organ pipe closed at one end and filled with carbon dioxide has the same frequency as that of the fundamental frequency of the open organic pipe. Calculate the length of each pipe. Given, velocity of sound in air is 330 m/s and in carbon dioxide is 270 m/s.

Solution: Fundamental frequency of open pipe containing air is

$$n = \frac{v}{2l}$$

Using $n = 460$ Hz and $v = 330$ m/s, we have

$$460 = \frac{330}{2l}$$

On solving this relation, we get $l = 0.359$ m. For the closed end pipe containing carbon dioxide, we have

$$n = \frac{v}{4l}$$

Using $n = 460$ Hz and $v = 270$ m/s, we have

$$460 = \frac{270}{4l}$$

On solving this relation, we get $l = 0.147$ m.

6. Interference of sound waves

When two or more sound waves travel through a common medium, superposition of the waves takes place. The particles of the medium oscillate according to the vector sum (resultant) of the displacements of the interfering waves at that instant. Owing to the superposition, the distribution of energy in the medium is different from that due to the individual waves separately. The resultant oscillation of a particle will depend on the amplitude, time-period, phase difference and direction of propagation of the interfering waves.

Let us consider two simple harmonic waves having the same frequency and same wavelength (and therefore the same velocity). Suppose, a and b be the amplitudes of two waves and ϕ the phase difference. In the discussion, it is assumed that the two waves have plane wavefront and propagate along a straight line. The displacements of a particle at an instant due to the waves are expressed as

$$y_1 = a \sin \frac{2\pi}{\lambda}(vt - x) \tag{6.21}$$

$$y_2 = b \sin\left[\frac{2\pi}{\lambda}(vt - x) + \phi\right] \tag{6.22}$$

Here, v is the velocity of propagation of each wave and λ the wave length. When the two waves are propagating simultaneously in a medium, the resultant displacement of a particle due to superposition of the waves is

$$y = y_1 + y_2 = a \sin\frac{2\pi}{\lambda}(vt - x) + b \sin\left[\frac{2\pi}{\lambda}(vt - x) + \phi\right]$$

$$= a \sin\frac{2\pi}{\lambda}(vt - x) + b\left[\sin\frac{2\pi}{\lambda}(vt - x) + \cos\phi + \cos\frac{2\pi}{\lambda}(vt - x) + \sin\phi\right]$$

$$= (a + b\cos\phi)\sin\frac{2\pi}{\lambda}(vt - x) + b\sin\phi\cos\frac{2\pi}{\lambda}(vt - x)$$

Since a, b and ϕ are constant, Let us assume

$$a + b\cos\phi = A\cos\theta \qquad \text{and} \qquad b\sin\phi = A\sin\theta$$

where A and θ are constants. Therefore,

$$y = A\cos\theta\sin\frac{2\pi}{\lambda}(vt - x) + A\sin\theta\cos\frac{2\pi}{\lambda}(vt - x)$$

$$= A\sin\left[\frac{2\pi}{\lambda}(vt - x) + \theta\right]$$

It shows that the resultant displacement has the same velocity and wavelength. The values of A and θ can be obtained as

$$A^2 = (a + b\cos\phi)^2 + (b\sin\phi)^2 = a^2 + b^2 + 2ab\cos\phi$$

or

$$A = \sqrt{a^2 + b^2 + 2ab\cos\phi} \qquad \text{and} \qquad \tan\theta = \frac{b\sin\phi}{a + b\cos\phi}$$

Spacial cases:

(i) When $\phi = 0,\ 2\pi,\ 4\pi,\ \ldots 2n\pi$, we have $\sin\phi = 0$ and $\cos\phi = 1$ and

$$A = \sqrt{a^2 + b^2 + 2ab} = a + b \qquad \text{and} \qquad \theta = 0$$

It is the case of constructive interference as shown in Figure 7. If $a = b$, we have $A = 2a$ and the intensity becomes 4 times of the intensity of individual waves.

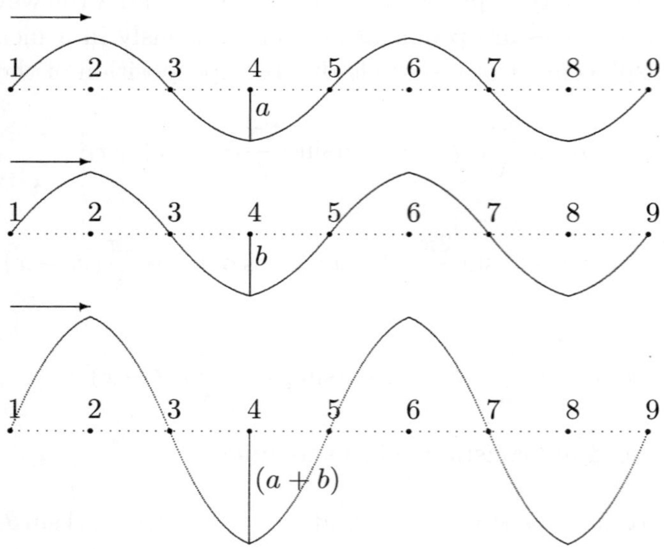

Figure 7: Constructive interference of two waves of amplitudes a and b, respectively.

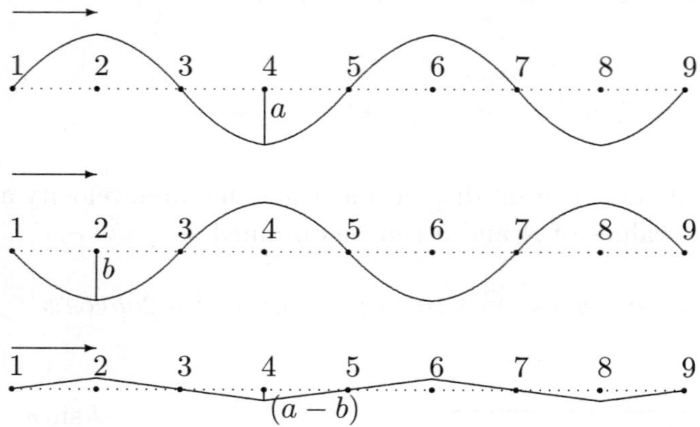

Figure 8: Destructive interference of two waves of amplitudes a and b, respectively.

(ii) When $\phi = \pi$, 3π, 5π, $\ldots (2n+1)\pi$, we have $\sin \phi = 0$ and $\cos \phi = -1$ and

$$A = \sqrt{a^2 + b^2 - 2ab} = |a - b| \qquad \text{and} \qquad \theta = 0 \quad \text{when} a \neq b$$

It is the case of destructive interference as shown in Figure 8. If $a = b$, we have $A = 0$ and the intensity becomes zero.

6.1 Conditions for interference of sound waves

(i) The two sources of sound must be coherent. They must produce waves of equal frequency. When the amplitudes of two waves are equal, the interference has good contrast.

(ii) The phase difference between the two waves must remain constant and should not change with time at any point.

(iii) The two waves should travel in the same direction.

6.2 Energy distribution due to interference of sound waves

For two waves expressed by (9) and (6.22), the resultant wave

$$y = A \sin\left[\frac{2\pi}{\lambda}(vt - x) + \theta\right]$$

where the amplitude A is

$$A = \sqrt{a^2 + b^2 + 2ab\cos\phi}$$

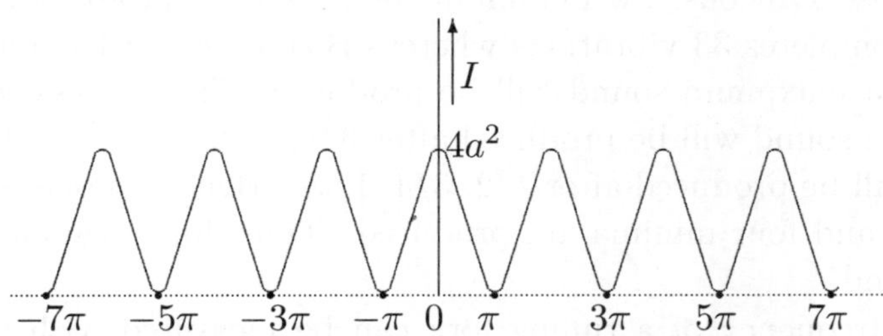

Figure 9: Variation of intensity.

When the amplitudes of two interfering waves are equal $(a = b)$, we have

$$A^2 = 2a^2(1 + \cos\phi) = 4a^2\cos^2(\phi/2)$$

This equation shows that the maximum intensity is $4a^2$ and the minimum intensity is zero, depending on the value of the phase difference ϕ between the interfering waves. Following the law of conservation of energy, the energy can neither be destroyed nor produced. Here, also the energy is not destroyed, but is redistributed from the position of minimum intensity to those of maximum intensity. For the maximum intensity positions, the intensity due to the two waves should be $2a^2$, but it is actually $4a^2$. As shown in Figure 9, the intensity varies from 0 to $4a^2$ and the average is again $2a^2$, which is equal to the uniform intensity.

Thus, the formation of maxima and minima due to the interference of two sound waves is in accordance with the law of conservation of energy.

7. Beats

When two sound waves of nearly equal frequency and amplitude interfere, the resultant wave consists of alternate maxima and minima. This phenomena in which waxing and vanning of sound at regular interval is heard is known as the beats. The number of beats heard per second is equal to the difference in frequencies of the two waves. However, the beats are heard only when the difference in the frequencies is not more than ten.

Let us look into the following situation. Consider two tuning forks A and B having frequencies 252 and 256, respectively. When the two tuning forks are sounded simultaneously, after 1/8 sec, the tuning fork A completes 31.5 vibrations whereas the tuning fork B completes 32 vibrations. Obviously, a minimum sound will be produced. After 1/4 sec, A completes 33 vibrations whereas B completes 64 vibrations. Obviously, a maximum sound will be produced. The process repeats and minimum sound will be produced after 3/8, 5/8, 7/8 sec and maximum sound will be produced after 1/2, 3/4, 1 sec. Hence, in one second, four maxima and four minima are produced. Thus, four beats are heard in one second.

The frequency of a tuning fork can be measured with the help of beats. Suppose, the frequency N of a tuning fork A is known. If the tuning fork A produces n beats with another tuning fork B. Then the frequency of the tuning fork B is $(N + n)$ or $(N - n)$. In order to decide among these two values. One of the two tuning forks A and B is loaded with wax or filed.

(i) After loading wax on a tuning fork, its frequency decreases.

(ii) After filing a tuning fork, its frequency increases.

Suppose, B is loaded with wax. If the number of beats decreases, the original frequency of B is $(N + n)$. When the number of beats increases, the original frequency of B is $(N - n)$. Suppose, A is loaded with wax. If the number of beats decreases, the original frequency of B is $(N - n)$. When the number of beats increases, the original frequency of B is $(N + n)$.

7.1 Analytical treatment of beats

Let us consider two sound waves of frequencies n_1 and n_2, where $(n_1 - n_2)$ is small. Suppose, a and b be the amplitudes of the waves, respectively. For the sake of simplicity, let us assume that the two waves are in phase at any point in the medium at $t = 0$. The displacement y_1 and y_2 due to the waves at any time t are respectively

$$y_1 = a \ \sin 2\pi n_1 t \qquad \text{and} \qquad y_2 = b \ \sin 2\pi n_2 t$$

The resultant displacement is

$$y = y_1 + y_2 = a \ \sin 2\pi n_1 t + b \ \sin 2\pi n_2 t$$

$$= a \ \sin 2\pi n_1 t + b \ \sin 2\pi [n_1 - (n_1 - n_2)]t$$

$$= a \ \sin 2\pi n_1 t + b \ \sin 2\pi n_1 t \ cos 2\pi (n_1 - n_2)t$$

$$-b \ \cos 2\pi n_1 t \ sin 2\pi (n_1 - n_2)t]$$

$$= \sin 2\pi n_1 t[a + b \ cos 2\pi (n_1 - n_2)t] - \cos 2\pi n_1 t[b \ sin 2\pi (n_1 - n_2)t]$$

Let us take

$$a + b \ cos 2\pi (n_1 - n_2)t = A \ \cos \theta$$

and

$$b \ sin 2\pi (n_1 - n_2)t = A \ \sin \theta$$

Thus, we have

$$y = \sin 2\pi n_1 t \ A \ \cos \theta - \cos 2\pi n_1 t \ A \ \sin \theta = A \ \sin(2\pi n_1 t - \theta)$$

where

$$A^2 = a^2 + b^2 + 2ab \ \cos 2\pi (n_1 - n_2)t$$

or

$$A = \sqrt{a^2 + b^2 + 2ab \ \cos 2\pi (n_1 - n_2)t} \qquad (6.23)$$

and

$$\tan\theta = \frac{b\,\sin 2\pi(n_1 - n_2)t}{a + b\,\cos 2\pi(n_1 - n_2)t} \tag{6.24}$$

From equations (6.23) and (6.24), it is obvious that the amplitude A and the phase angle θ vary with time. Let us consider various situations.

(i) When $(n_1 - n_2)t = K$ is an integer, we have $\sin 2\pi K = 0$ and $\cos 2\pi K = 1$. Then we have

$$A = \sqrt{a^2 + b^2 + 2ab} = a + b \qquad \text{and} \qquad \theta = 0$$

It is the case of constructive interference. Hence, the amplitude is the maximum when

$$(n_1 - n_2)t = K \qquad \text{or} \qquad t = \frac{K}{(n_1 - n_2)}$$

Thus, the amplitude is the maximum at the times

$$0, \qquad \frac{1}{(n_1 - n_2)}, \qquad \frac{2}{(n_1 - n_2)}, \qquad \frac{3}{(n_1 - n_2)}, \qquad \cdots$$

As the intensity of sound is directly proportional to the square of the amplitude, the maximum intensity of sound will be heard at the times

$$0, \qquad \frac{1}{(n_1 - n_2)}, \qquad \frac{2}{(n_1 - n_2)}, \qquad \frac{3}{(n_1 - n_2)}, \qquad \cdots$$

(ii) When $2(n_1 - n_2)t = (2K + 1)$, where K is an integer, we have $\sin(2K + 1)\pi = 0$ and $\cos(2K + 1)\pi = -1$. Then we have

$$A = \sqrt{a^2 + b^2 - 2ab} = |a - b| \qquad \text{and} \qquad \theta = 0 \qquad \text{when } a \neq b$$

It is the case of destructive interference. Hence, the amplitude is the minimum when

$$2(n_1 - n_2)t = (2K + 1) \qquad \text{or} \qquad t = \frac{(2K + 1)}{2(n_1 - n_2)}$$

Thus, the amplitude is the minimum at the times

$$\frac{1}{2(n_1 - n_2)}, \qquad \frac{3}{2(n_1 - n_2)}, \qquad \frac{5}{2(n_1 - n_2)}, \qquad \frac{7}{2(n_1 - n_2)}, \qquad \cdots$$

The minimum intensity of sound will be heard at the times

$$\frac{1}{2(n_1 - n_2)}, \qquad \frac{3}{2(n_1 - n_2)}, \qquad \frac{5}{2(n_1 - n_2)}, \qquad \frac{7}{2(n_1 - n_2)}, \qquad \cdots$$

Obviously, the maximum and minimum occur alternatively after equal intervals of time $1/2(n_1 - n_2)$. The time interval between two successive maxima or two minima is $1/(n_1 - n_2)$. Thus, the number of beats produced per second is

$$= \frac{1}{1/(n_1 - n_2)} = (n_1 - n_2)$$

It shows that the number of beats per second is equal to the difference between the frequencies of two waves. When the amplitudes of two waves are equal ($a = b$), the maximum amplitude of the resultant is $2a$ and the minimum amplitude is zero. Thus, the positions of minima will have zero intensity.

Exercise 11: A tuning fork A produces 6 beats per second with other tuning fork B whose frequency is 250. The tuning fork A is filed and the beats occur at the shorter intervals. Calculate the original frequency of A.

Solution: The frequency of tuning fork B is 250

The number of beats per second is 6.

The frequency of tuning fork A is either 250 + 6 = 256 or 250 - 6 = 244

When the tuning fork A is filed its frequency increases. On filing A, the beats occur at the shorter intervals, *i.e.*, the number of beats increases. Hence, the increased frequency of A must go away from 250. It can only be possible when the original frequency of A is 256.

Exercise 12: A tuning fork A produces 6 beats per second with the tuning fork B whose frequency is 350 and 8 beats per second with the tuning fork C having frequency 364. Find the frequency of A.

Solution: The frequency of tuning fork B is 350

The number of beats per second is 6.

The frequency of tuning fork A is either 350 + 6 = 356 or 350 - 6 = 344

The frequency of tuning fork C is 364

The number of beats per second is 8.

The frequency of tuning fork A is either 364 + 8 = 372 or 364 - 8 = 356

Thus, the frequency of tuning fork A is 356.

Exercise 13: Two tuning forks A and B produce 6 beats per second. The frequency of the tuning fork A is 356. The tuning fork B is filed.

Two tuning forks A and B again produced 6 beats per second. Calculate the frequency of B before and after filing.

Solution: The frequency of the tuning fork A is 356.

The number of beats is 6.

The frequency of B is either $356 + 6 = 362$ or $356 - 6 = 350$.

After filing of B, the frequency of B increases. Since again 6 beats per second are produced, the original frequency of B is 350. After filing, the frequency of B is 362.

Exercise 14: A tuning fork A having frequency 380 Hz gives 8 beats per second when sounded with other tuning fork B. On loading B with a little wax, the number of beats per second becomes 4. Calculate the original frequency of B.

Solution: The frequency of the tuning fork A is 380.

The number of beats is 8.

The frequency of B is either $380 + 8 = 388$ or $380 - 8 = 372$.

On loading little wax on B, its frequency decreases. Now, the number of beats per second becomes 4. Therefore, the original frequency of B is 388.

8. Intensity of sound

The intensity of sound is defined as the average transfer of energy per unit area per unit time, where the area is perpendicular to the direction of propagation of the sound. In the last chapter, we have seen when a particle of mass m executes a simple harmonic motion with amplitude A and frequency n, the energy E of the particle is

$$E = 2\pi^2 mn^2 A^2$$

For the velocity v, the amount of energy transfer per unit area per second is

$$I = 2\pi^2 \rho\, n^2 A^2 v \tag{6.25}$$

where ρ is the density of the medium through which the sound wave is propagating and the particles of the medium are executing simple harmonic motion. The velocity v of the sound wave is

$$v = \sqrt{\frac{E}{\rho}}$$

The sound wave propagates through the medium in the longitudinal mode. For a longitudinal wave

$$y = A \sin \frac{2\pi}{\lambda}(vt - x) \qquad (6.26)$$

propagating in a gas, the volume strain[1] is dy/dx and the modulus of elasticity E of the medium is

$$E = -\frac{p}{dy/dx}$$

---

[1]Let us consider a plane wave propagating through a gas. It can be noted that a wave can propagate through a gas in the longitudinal mode. The particles in the medium execute simple harmonic motion along the direction of propagation of the wave. Let us consider a wave propagating along the positive direction of x-axis. Imagine a cylindrical gas column along x axis.

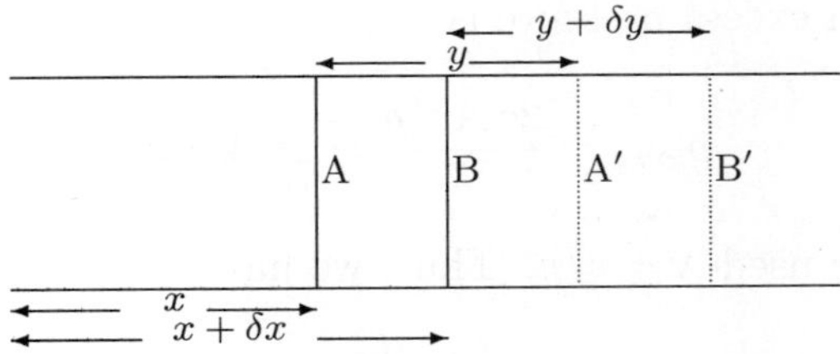

Now consider in equilibrium, two planes A and B (very close to each other) of the gas column situated at the positions x and $x + \delta x$, respectively, from the origin, as shown in the figure above. When the wave propagates, at some instant, the particles of plane A are displaced by an amount y to the position A$'$ and those of plane B are displaced by an amount $y + \delta y$ to the position B$'$. Since dy/dx is the rate of change of displacement with distance, we have

$$y + \delta y = y + \left(\frac{dy}{dx}\right)\delta x$$

Between the planes A and B, the initial volume of the gas column is

$$= \alpha(\text{AB}) = \alpha\delta x$$

The final volume of the gas between the planes A$'$ and B$'$ is

$$= \alpha(\text{A}'\text{B}') = \alpha(\delta x + \delta y) = \alpha\left[\delta x + \left(\frac{dy}{dx}\right)\delta x\right]$$

The change is volume is

$$= \alpha\left[\delta x + \left(\frac{dy}{dx}\right)\delta x\right] - \alpha\delta x = \alpha\left(\frac{dy}{dx}\right)\delta x$$

The volume strain is

$$\text{volume strain} = \frac{\text{change in volume}}{\text{initial volume}} = \frac{\alpha\,(dy/dx)\,\delta x}{\alpha\,\delta x} = \frac{dy}{dx}$$

where p is the excess of pressure applied. Therefore, the velocity of the wave is

$$v = \sqrt{-\frac{p}{\rho \, dy/dx}} \qquad \text{or} \qquad p = -v^2 \rho \, \frac{dy}{dx} \qquad (6.27)$$

On differentiation of equation (6.26) with respect to x, we have A simple harmonic wave is represented by

$$\frac{dy}{dx} = -\frac{2\pi A}{\lambda} \cos \frac{2\pi}{\lambda}(vt - x)$$

Substituting this value in equation (6.27), we get

$$p = \frac{2\pi A v^2 \rho}{\lambda} \cos \frac{2\pi}{\lambda}(vt - x)$$

The maximum excess pressure is

$$p_{max} = \frac{2\pi A v^2 \rho}{\lambda} = 2\pi A v n \rho \qquad (6.28)$$

where we have used $\lambda = v/n$. Thus, we have

$$p = p_{max} \cos \frac{2\pi}{\lambda}(vt - x)$$

From equations (6.25) and (6.28), we have

$$I = \frac{p_{max}^2}{2\rho v}$$

This equation shows that the intensity of sound varies directly as the square of the excess pressure.

9. Measurement of intensity of sound

The intensity of sound is defined as the quantity of energy propagating through a unit area per unit time, the direction of propagation being perpendicular to the area. The unit of intensity in the SI system is J/m^2. The amount of power transmitted per unit area is measured in watts/m^2.

According to the Weber-Fecher law, the loudness of sound produced is proportional to the logarithm of intensity. For the intensity I, the loudness S is

$$S = K \ln I$$

where K is a constant. Thus,

$$\frac{\mathrm{d}S}{\mathrm{d}I} = \frac{K}{I}$$

The quantity $\mathrm{d}S/\mathrm{d}I$ is known as the sensitiveness of the ear. Thus, the sensitiveness of the ear decreases with the increase of the intensity of the sound.

For all practical measurements, the relative intensity is important and not the absolute intensity. Thus, the intensity of sound is measured as its ratio to the standard intensity I_0. We define the intensity level of sound as

$$\text{intensity level} = \frac{I}{I_0}$$

A person speaking in a normal conversational tone emits energy at the rate of about 10^{-5} J/s. The mouth aperture at the time of normal conversation is about 10^{-3} m^2. Thus, the intensity of sound is $10^{-5}/10^{-3}$ m^2 = 10^{-2} J/s m^2 = 10^{-2} watt/m^2. This value of intensity is defined as the standard intensity and is denoted by I_0. When a person shouts, the intensity may be up to $100\, I_0 = 1$ watt/m^2. When the intensity is from $100\, I_0$ to $1000\, I_0$, the listener feels pain.

The faintest sound that can be heard besides its intensity depends also on the frequency of the wave. The average person's threshold of audibility is about $10^{-10} I_0$ for a frequency of 440 Hz. Thus, the range of hearing for a human ear is from $10^{-10} I_0$ to $100\, I_0$. That is, the human ear has a dynamic range of 10^{12} in the intensity.

9.1 Decibel (dB)

When the intensity of sound increases by a factor of 10, the increase in intensity is said to be 1 bel. Hence, the dynamic range of audibility of the human ear is 12 bels or 120 decibels.

Consider the loudness S and S_0 corresponding to the intensities I and I_0, respectively. We have

$$S = K\, \log I \qquad \text{and} \qquad S_0 = K\, \log I_0$$

Notice that here we have taken log (logarithm base 10) instead of ln (logarithm base e) and the constant K is modified accordingly. The intensity level L is the difference in loudness. Thus, we have

$$L = S - S_0 = K\, \log I - K\, \log I_0 = K\, \log\!\left(\frac{I}{I_0}\right)$$

For $K = 1$, we have

$$L = \log\left(\frac{I}{I_0}\right)$$

Here, the unit of L is bel. The L is generally expressed in the decibels (dB) and we have

$$L = 10 \log\left(\frac{I}{I_0}\right)$$

For $L = 1$ dB, we have

$$1 = 10 \log\left(\frac{I}{I_0}\right) \qquad \text{or} \qquad \frac{I}{I_0} = 10^{0.1} = 1.26$$

It shows that the energy level changes by 1 decibel when the intensity of sound changes by 26%. The lowest change in the intensity level that can be detected by a human ear is 1 dB.

Exercise 15: calculate the change in the intensity level when the intensity of sound increases by 100 times its original intensity.

Solution: For the original intensity I_0 and final intensity I, we are given

$$\frac{I}{I_0} = 100$$

The increase in intensity level L is

$$L = 10 \log\left(\frac{I}{I_0}\right) = 10 \log 100 = 10(2) = 20 \text{ dB}$$

10. Doppler effect

A person can easily experience that the pitch (frequency) of a note (sound wave) appears to change when either the source producing the wave or the observer is moving relative to each other. When the source approaches the observer or both the source and observer approach to each other, the frequency appears to increase as compared to the actual frequency produced by the source. Similarly, when the source moves away from the observer or both the source and observer move away from each other, the frequency appears to decrease as compared to the actual frequency produced by the source.

It can be elaborated as the following. Suppose a passenger is standing on a platform and waiting for the arrival of the train. The apparent frequency of the whistle appears to increase as compared to the original frequency when the engine is approaching the passenger. When the engine passes the passenger and moves away from the passenger, the apparent frequency of the whistle of the engine appears to decrease as compared to the original frequency. This apparent change in the frequency due to the relative motion between the source and the observer is known as the Doppler effect.

Let us now consider various situations for relative motions of the source and the observer as the following.

A. Observer is at rest and source is in motion

(i) Suppose a source S is producing sound of pitch n and wavelength λ. Let the velocity of sound is v and the source moves with a velocity a towards the observer O.

In one second n waves will be adjusted within the length of $(v - a)$ and the apparent wavelength is

$$\lambda' = \frac{v - a}{n}$$

Therefore, the apparent frequency is

$$n' = \frac{v}{\lambda'} = \frac{v}{v - a}\, n$$

Hence, the apparent frequency of the wave increases when the source moves towards the stationary observer.

(ii) Suppose the source S moves with a velocity a away from the observer O.

In one second n waves will be adjusted within the length of $(v + a)$ and the apparent wavelength is

$$\lambda' = \frac{v + a}{n}$$

Therefore, the apparent frequency is

$$n' = \frac{v}{\lambda'} = \frac{v}{v + a}\, n$$

Hence, the apparent frequency of the wave decreases when the source moves away from the stationary observer.

B. Source is at rest and observer is in motion

(i) Suppose a source S is producing sound of pitch n and wavelength λ is stationary. Let the velocity of sound is v and the observer O moves with a velocity b towards the source.

In this case, the observer receives more number of waves in one second. The apparent wavelength remains the same and the apparent frequency is

$$n' = n + \frac{b}{\lambda}$$

Using $\lambda = v/n$, we get

$$n' = n + \frac{b}{v/n} = \left(\frac{v + b}{v}\right) n$$

Hence, the apparent frequency of the wave increases when the observer moves towards the stationary source.

(ii) Suppose the observer O moves with the velocity b away from the source S.

In this case, the observer receives less number of waves in one second. The apparent wavelength remains the same and the apparent frequency is

$$n' = n - \frac{b}{\lambda}$$

Using $\lambda = v/n$, we get

$$n' = n - \frac{b}{v/n} = \left(\frac{v - b}{v}\right) n$$

Hence, the apparent frequency of the wave decreases when the observer moves away from the stationary source.

C. Both the source and observer are in motion

Suppose a source S, producing a sound wave of frequency n and wavelength λ moves towards the observer O with velocity a and the observer moves away from the source with velocity b. Consider that the velocity of the sound wave is v.

When the source moves towards the observer with velocity a and the observer is moving away from the source with velocity b, then we have

$$\lambda' = \frac{v - a}{n}$$

and

$$n' = \frac{v - b}{\lambda'}$$

Using the value of λ' here, we get

$$n' = \left(\frac{v - b}{v - a}\right) n \tag{6.29}$$

Special cases

(i) **The source and observer are moving towards each other.**

Then in equation (6.29), the velocity b is taken as negative and we have

$$n' = \left(\frac{v - [-b]}{v - a}\right) n = \left(\frac{v + b}{v - a}\right) n$$

(ii) **The source and observer are moving away from each other.**

Then in equation (6.29), the velocity a is taken as negative and we have

$$n' = \left(\frac{v-b}{v-[-a]}\right) n = \left(\frac{v-b}{v+a}\right) n$$

(iii) **The source is moving away the observer and the observer is moving towards the source.**

Then in equation (6.29), both the velocities a and b are taken as negative and we have

$$n' = \left(\frac{v-[-b]}{v-[-a]}\right) n = \left(\frac{v+b}{v+a}\right) n$$

Note: While solving numerical problems, one should keep in mind only one formula. For example, one can take the formula

$$n' = \left(\frac{v-b}{v-a}\right) n$$

where the source moves towards the observer and observer moves away from the source. Then the signs of the velocities of the source and the observer can be changed accordingly.

D. Effect of the velocity of medium

Suppose the medium through which the wave is propagating is moving with velocity w in the direction of propagation of the wave. The apparent velocity of the wave will be $(v + w)$. Thus, in all the above relations, we have to change v by $(v + w)$. When the medium is moving in the direction opposite to the direction of propagation of the wave, the sign of w is to be changed.

Thus, for example, for the apparent frequency, one can take the formula

$$n' = \left(\frac{v+w-b}{v+w-a}\right) n$$

where the source moves towards the observer, the observer moves away from the source, and the medium moves in the direction of the propagation of the wave. It can be used as a general expression. Then the signs of the velocities of source, observer and medium can be changed accordingly.

Exercise 16: A passenger is standing on a platform and waiting for the arrival of the train. An engine while approaching the passenger blows a whistle of frequency 620 Hz. The speed of the engine is 72 km/hr and the velocity of sound is 330 m/s. Calculate the apparent frequency of the engine as heard by the passenger.

Solution: We have $v = 330$ m/s, $a = 72$ km/hr $= 20$ m/s and $n = 640$ Hz. The apparent frequency of the source heard by the passenger is

$$n' = \left(\frac{v}{v-a}\right) n = \left(\frac{330}{330-20}\right) 620 = 660 \text{ Hz}$$

Exercise 17: A passenger is standing on a platform and waiting for the arrival of the train. An engine while moving away from the passenger with the speed of 72 km/hr blows a whistle of frequency 680 Hz. The velocity of sound is 320 m/s. Calculate the apparent frequency of the engine as heard by the passenger.

Solution: We have $v = 320$ m/s, $a = 72$ km/hr $= 20$ m/s and $n = 680$ Hz. Thus, the apparent frequency of the source heard by the passenger is

$$n' = \left(\frac{v}{v+a}\right) n = \left(\frac{320}{320+20}\right) 680 = 640 \text{ Hz}$$

Exercise 18: Two airplanes A and B are approaching each other with a speed of 360 km/hr. The frequency of whistle emitted by A is 940 Hz. Calculate the apparent frequency of the whistle heard by the passenger of the airplane B. Velocity of sound in air is 360 m/s.

Solution: We have $n = 940$ Hz, $v = 360$ m/s, $a = 360$ km/hr $= 100$ m/s and $b = 360$ km/hr $= 100$ m/s. As the source and observer are moving towards each other, the apparent frequency is

$$n' = \left(\frac{v+b}{v-a}\right) n = \left(\frac{360+100}{360-100}\right) 940 = 1663 \text{ Hz}$$

Exercise 19: Two airplanes A and B are moving away from each other with a speed of 360 km/hr. The frequency of whistle emitted by A is 980 Hz. Calculate the apparent frequency of the whistle heard by the passenger of the airplane B. Velocity of sound in air is 370 m/s.

Solution: We have $n = 980$ Hz, $v = 370$ m/s, $a = 360$ km/hr $= 100$ m/s and $b = 360$ km/hr $= 100$ m/s. As the source and observer are moving away from each other, the apparent frequency is

$$n' = \left(\frac{v-b}{v+a}\right) n = \left(\frac{370-100}{370+100}\right) 980 = 563 \text{ Hz}$$

Exercise 20: Two airplanes A and B pass each other in opposite directions one of then is blowing a whistle of frequency 980 Hz. Calculate the frequencies of the notes heard in the other airplane (i) before and (ii) after the airplanes have passed each other. Velocity of either of the airplane is 720 km/hr and the velocity of sound in air is 430 m/s.

Solution: We have $n = 980$ Hz, $v = 430$ m/s, $a = 720$ km/hr $= 200$ m/s and $b = 720$ km/hr $= 200$ m/s.

(i) The source and observer are moving towards each other, the apparent frequency is

$$n' = \left(\frac{v+b}{v-a}\right) n = \left(\frac{430+200}{430-200}\right) 980 = 2684 \text{ Hz}$$

(ii) The source and observer are moving away from each other, the apparent frequency is

$$n' = \left(\frac{v-b}{v+a}\right) n = \left(\frac{430-200}{430+200}\right) 980 = 578 \text{ Hz}$$

Exercise 21: A passenger standing on a railway platform observed that as a train passes through the station at 90 km/hr, the frequency of the whistle appeared to drop by 400 Hz. Calculate the frequency of the whistle. Velocity of sound in air is 410 m/s.

Solution: We have $v = 410$ m/s, $a = 90$ km/hr $= 25$ m/s and $b = 0$ m/s.

(i) When the source is moving towards the observer, the apparent frequency is

$$n_1 = \left(\frac{v}{v-a}\right) n$$

(ii) When the source is moving away from the observer, the apparent frequency is

$$n_2 = \left(\frac{v}{v+a}\right) n$$

The difference in the frequencies is

$$n_1 - n_2 = n \left[\left(\frac{v}{v-a}\right) - \left(\frac{v}{v+a}\right)\right]$$

Using the values of various physical quantities, we have

$$400 = n \left[\left(\frac{410}{410-25}\right) - \left(\frac{410}{410+25}\right)\right]$$

On solving, we get $n = 3268$ Hz

Exercise 22: A car producing sound of frequency 250 Hz is moving away from a stationary observer and towards a wall with a velocity of 72 km/hr. Calculate the number of beats heard per second by the observer. Velocity of sound is 410 m/s.

Solution: We have $n = 250$ Hz, $v = 410$ m/s and velocity of car $a = 72$ km/hr $= 20$ m/s. The observer hears the direct sound from the car. The apparent frequency is

$$n_1 = \left(\frac{v}{v+a}\right) n = \left(\frac{410}{410+20}\right) 250 = 238.4 \text{ Hz}$$

The observer hears also the sound reflected by the wall. Here, the source is moving towards the observer and the apparent frequency is

$$n_2 = \left(\frac{v}{v-a}\right) n = \left(\frac{410}{410-20}\right) 250 = 262.8 \text{ Hz}$$

Thus, the number of beats produced in one second is

$$n_2 - n_1 = 262.8 - 238.4 = 24.4$$

Exercise 23: Apparent frequency of whistle of an engine changes in the ratio 5 : 4 when the engine passes a stationary passenger standing on a platform. If the velocity of the sound is 380 m/s, calculate the velocity of the engine.

Solution: We have $v = 380$ m/s. Let the original frequency of the whistle is n and the engine is moving with velocity a.

(i) When the engine moves towards the stationary passenger the apparent frequency is

$$n_1 = \left(\frac{v}{v-a}\right) n \tag{6.30}$$

(ii) When the engine moves away from the stationary passenger the apparent frequency is

$$n_2 = \left(\frac{v}{v+a}\right) n \tag{6.31}$$

On dividing equation (6.30) by (6.33), we have

$$\frac{n_1}{n_2} = \left(\frac{v+a}{v-a}\right) \tag{6.32}$$

Using $n_1/n_2 = 5/4$ and the value of v, we get

$$\frac{5}{4} = \left(\frac{380 + a}{380 - a}\right) \tag{6.33}$$

On solving, we get $a = 42.2$ m/s.

Exercise 24: Two trains moving in opposite directions with 72 km/hr each, cross each other while one of them is whistling. If the frequency of the sound produced by the whistle is 750 Hz, calculate the apparent frequency as heard by a passenger sitting in the other train: (i) before the trains cross each other, and (ii) after the trains cross each other. velocity of sound in air is $v = 350$ m/s.

Solution: Given $a = 72$ km/hr $= 20$ m/s, $b = 72$ km/hr $= 20$ m/s, $n = 750$ Hz.

(i) Before the trains cross each other. The trains are moving towards each other. The apparent frequency is

$$n_1 = \left(\frac{v + b}{v - a}\right) n = \left(\frac{350 + 20}{350 - 20}\right) 750 = 840.9 \text{ Hz}$$

(ii) After the trains cross each other. The trains are moving away from each other. The apparent frequency is

$$n_2 = \left(\frac{v - b}{v + a}\right) n = \left(\frac{350 - 20}{350 + 20}\right) 750 = 668.9 \text{ Hz}$$

Exercise 25: A train fitted with two sounding horns which have difference in frequencies by 180 Hz is moving with the speed of 72 km/hr towards a standing passenger. Calculate the difference in the frequencies of the sounds heard by the passenger. Velocity of sound in air $= 360$ m/s.

Solution: Suppose, n_1 and n_2 are two frequencies produced by the horns fitted in the train. We have $n_1 - n_2 = 180$ Hz. We have $v = 360$ m/s, $a = 72$ km/hr $= 20$ m/s and $b = 0$. The apparent frequencies are

$$n_1' = \left(\frac{v}{v - a}\right) n_1 \qquad \text{and} \qquad n_2' = \left(\frac{v}{v - a}\right) n_2$$

Therefore, the difference between the apparent frequencies is

$$n_1' - n_2' = \left(\frac{v}{v - a}\right) (n_1 - n_2)$$

Using the values, we have

$$n_1' - n_2' = \left(\frac{360}{360 - 20}\right) 180 = 190.6 \text{ Hz}$$

11. Doppler effect in light

There is change in the frequency of radiation (light) when the source or the observer moves with respect to each other. This phenomenon is known as the Doppler effect. It is similar to the apparent change in the frequency of sound when either the source or the observer, or both are in motion with respect to each other. However, there is difference between the sound (mechanics) waves and the light waves. In case of the light waves, any material medium is not necessary whereas in case of sound waves some material medium is necessary. The light waves have the largest speed in the vacuum

In case of the sound waves, the Doppler effect is asymmetric, *i.e.*, the apparent frequencies are different when the source moves towards a stationary observer and when the observer moves towards a stationary source. Suppose a source S is moving with velocity a towards a stationary observer O.

Then the apparent frequency is

$$n' = \frac{v}{v - a}\, n \tag{6.34}$$

Suppose an observer O is moving with velocity a towards a stationary source S.

Then the apparent frequency is

$$n' = \left(\frac{v + a}{v}\right) n \tag{6.35}$$

where n is the original frequency and v the velocity of the sound wave. Comparison of equations (6.34) and (6.35) shows that the Doppler effect in the sound waves is asymmetric.

But, the asymmetry is not applicable in case of the light waves. In case of light waves, the Doppler effect is symmetric. That is, the apparent frequency is the same when, for example, either the observer is moving away from the source or the source is moving away from the observer.

(i) Suppose an observer is stationary and a source is moving towards the observer with a velocity v. Let the velocity of light be c. For frequency ν and wavelength λ, we have

$$c = \nu\lambda \tag{6.36}$$

As there is no medium, the observer will receive more waves due to the motion of the source. The wavelength does not change. The apparent frequency is

$$\nu' = \nu + \frac{v}{\lambda} \tag{6.37}$$

Using the value of λ from equation (6.36) in (6.37), we have

$$\nu' = \nu + \frac{v}{c/\nu} \qquad \text{or} \qquad \nu' = \nu\left(1 + \frac{v}{c}\right) \tag{6.38}$$

(ii) Suppose a source is stationary and an observer is moving towards the observer with a velocity v.

As there is no medium, the observer will receive more waves due to its own motion. The wavelength does not change. The apparent frequency is

$$\nu' = \nu + \frac{v}{\lambda} \tag{6.39}$$

Using the value of λ from equation (6.36) in (6.39), we have

$$\nu' = \nu + \frac{v}{c/\nu} \qquad \text{or} \qquad \nu' = \nu\left(1 + \frac{v}{c}\right) \tag{6.40}$$

Comparison of equations (6.38) and (6.40) shows that the Doppler effect in light is symmetric.

(iii) Suppose an observer is stationary and a source is moving away from the observer with a velocity v.

As there is no medium, the observer will receive less waves due to the motion of the source. The wavelength does not change. The apparent frequency is

$$\nu' = \nu - \frac{v}{\lambda} \tag{6.41}$$

Using the value of λ from equation (6.36) in (6.41), we have

$$\nu' = \nu - \frac{v}{c/\nu} \qquad \text{or} \qquad \nu' = \nu\left(1 - \frac{v}{c}\right) \tag{6.42}$$

(iv) Suppose a source is stationary and an observer is moving away from the observer with a velocity v.

As there is no medium, the observer will receive less waves due to its own motion. The wavelength does not change. The apparent frequency is

$$\nu' = \nu - \frac{v}{\lambda} \tag{6.43}$$

Using the value of λ from equation (6.36) in (6.43), we have

$$\nu' = \nu - \frac{v}{c/\nu} \qquad \text{or} \qquad \nu' = \nu\left(1 - \frac{v}{c}\right) \tag{6.44}$$

Comparison of equations (6.47) and (6.44) shows that the Doppler effect in light is symmetric.

(v) Suppose a source and an observer are moving towards each other and each of them is moving with velocity v.

The apparent frequency is

$$\nu' = \nu\left(1 + \frac{v}{c}\right)^2 \tag{6.45}$$

(vi) Suppose a source and an observer are moving away from each other and each of them is moving with velocity v.

$$\underset{\bullet}{\text{S}} \quad v \quad \xrightarrow{\quad c \quad} \quad v \quad \underset{\bullet}{\text{O}}$$

The apparent frequency is

$$\nu' = \nu\left(1 - \frac{v}{c}\right)^2 \tag{6.46}$$

(vii) The Doppler effect in light is of great importance. In the context of light from the astronomical objects, an observer is situated on the surface of the earth and, in the scenario of the expanding universe, the objects is always moving away from the observer. Then the apparent frequency is

$$\nu' = \nu\left(1 - \frac{v}{c}\right) \tag{6.47}$$

Using the relations

$$\lambda = \frac{c}{\nu} \qquad \text{and} \qquad \lambda' = \frac{c}{\nu'}$$

we get

$$\lambda' = \lambda\left(1 - \frac{v}{c}\right)^{-1} = \lambda\left(1 + \frac{v}{c}\right)$$

It shows that the apparent frequency from an astronomical object is always less than the original frequency. Consequently, the radiations from the astronomical objects always shows the red-shift.

Exercise 26: The wavelength of an atomic transition observed in a laboratory is 5800 Å. When this transition was found from a cosmic object, its wavelength is found to be 5810 Å. Calculate the speed of the cosmic object.

Solution: Given, $\lambda = 5800$ Å, $\lambda' = 5810$ Å, speed of light $c = 3 \times 10^8$ m/s. Using the values, we have

$$5810 = 5800\left(1 + \frac{v}{3 \times 10^8}\right)$$

Therefore,

$$\frac{5810}{5800} = 1 + \frac{v}{3 \times 10^8} \qquad \text{or} \qquad \frac{5810 - 5800}{5800} = \frac{v}{3 \times 10^8}$$

On solving this relation, we get $v = 5.17 \times 10^5$ m/s.

11.1 Applications

Though the Doppler effect is equally important in case of both the sound waves and the light waves, the Doppler effect in light can be seen easily with the help of natural phenomena. Here, we shall discuss about some application of Doppler effect in light.

(i) **Velocity of rotation of the sun.** The sun is rotating about its own axis. From the surface of the earth, the two sides (west and east edges) of the sun appear to move in opposite directions. One side moves away from the observer situated on the earth's surface whereas the other side moves towards the observer. By the study of Doppler shift from the light received the west and east edges, it has been found that the shift is due to a velocity of about 2 km/s for the rotation of the sun. Moreover, no such shift has been found from the light received from the north and south edges. It shows that the sun rotates about the north-south axis.

(ii) **Discovery of binary stars.** Observations from the sky shows that some stars are in the form of binary stars (a pair of stars) which revolve about an axis passing through their center-of-mass. When one of them is moving towards the earth, the other is moving away from the earth. Thus, there is shift in their spectral lines and a single line is split into two lines whose separation depends on the time and the time-period of revolution of binary system. By using this technique, a number of binary stars have been discovered.

(iii) **Red shift.** In the scenario of the expanding universe, the radiations from the distant astronomical objects always show the red-shift.

(iv) **Rings around the Saturn.** The planet Saturn has been found surrounded by concentric rings. With the help of the Doppler effect, it has been found that these rings are not solids but have a number of satellites revolving around the Saturn. If the rings have been solids, the outer edge of the ring should have larger velocity than the inner edge. Following the principle of satellite, we have

$$\frac{mv^2}{R} = \frac{GMm}{R^2} \qquad \text{or} \qquad v = \sqrt{\frac{GM}{R}}$$

where m and v are the mass and velocity, respectively, of the satellite; M is the mass of Saturn and R the mean distance between the satellite and Saturn. This relation shows that the velocity of the satellite in the inner orbit is more than in the outer orbit. This fact has been found with the help of Doppler effect. Hence, the rings of Saturn are not solids but have a number of satellites revolving around it.

12. Supersonic speeds

A body moving with a speed larger than that of the sound is said to be moving with the supersonic speed. Motion of a bullet fired by a gun, speed of a fighter jet are examples of bodies moving with the supersonic speeds. The ratio of the velocity of a wave to the speed of sound is defined as the Mach number. For a supersonic speed obviously the Mach number is larger than 1. The nature of the wavefront for the supersonic speed is conical in shape.

Suppose a body is moving at supersonic speed V and the velocity of sound is v, as shown in Figure 10. Let P be the position of the body at a time, taken as the initial time. Suppose after the time t, the body reaches the point Q, so that the distance PQ $= Vt$. In the same time t, the longitudinal compressional waves of sound produced at the point P travel at distance vt, producing a spherical wavefront of radius vt. From the point Q, draw the tangents QA and QB on the sphere. Then, PA $=$ PB $= vt$. Since $V > v$, we have PQ $>$ PA.

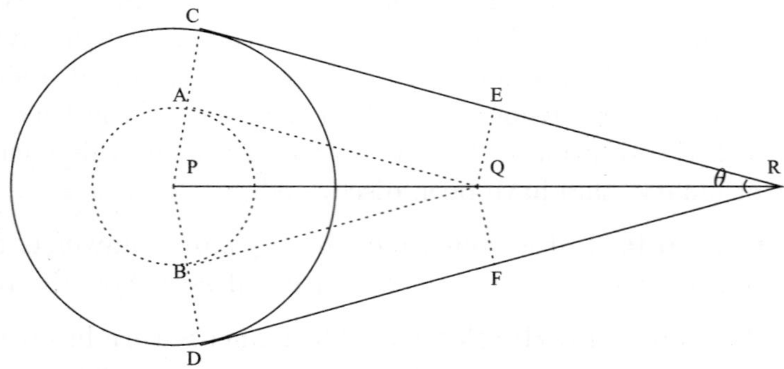

Figure 10: Illustration of supersonic speeds.

After a time $2t$, the body reaches the point R. From Q to R, the time taken is t, and therefore QR $= Vt$. The sound waves produced at P will construct a sphere of radius $2vt$. From the point R, draw the tangents RC and RD on this sphere. Then, PC $=$ PD $= 2vt$. From the point Q, draw perpendiculars QE and QF on the tangents RC and RD, respectively.

If we look this phenomenon in other way that a supersonic wave is produced at the point R. That will be at the point Q in time t and the point P in time $2t$. Since the lines RC and RD form a cone with the apex at the point R. It shows that the wavefront is conical in shape. The cone angle θ obviously depends on the speed of sound and the speed of the body in the medium.

In the $\triangle$ RPC and $\triangle$ RQE:

$\angle$ CRP is common

$\angle$ REQ = $\angle$ RCP = $90°$

Thus, the $\triangle$ RPC and $\triangle$ RQE are similar triangles. Since

$$RP = 2\,RQ \qquad \text{therefore,} \qquad PC = 2\,QE$$

Hence, we have

$$\sin\theta/2 = \frac{PC}{PR} = \frac{2vt}{2Vt} = \frac{v}{V} \qquad \text{or} \qquad \theta = 2\left(\sin^{-1}\frac{v}{V}\right)$$

This equation shows that the cone angle decreases with the increase of the speed of the body.

12.1 Shock waves

A shock wave is a disturbance propagating with the supersonic speed. It is characterized by an abrupt, nearly discontinuous change in the characteristics of the medium. Across the shock there is always an extremely rapid rise in pressure, temperature and density of the flow. It is one of the different ways in which a gas in a supersonic flow can be compressed. The method of compression of a gas results in higher temperature and higher density for a given pressure ratio. Like an ordinary wave, it carries energy and can propagate through a medium (solid, liquid, gas or plasma) or in absence of a material medium, through a field such as the electromagnetic field. The energy of a shock wave dissipates relatively quickly with distance. Thus, a sonic boom associated with the passage of a supersonic aircraft is the sound wave resulting from the degradation and merging of the the shock wave and the expansion produced by the aircraft. When a shock wave passes through matter, total energy is preserved but the energy which can be extracted as work decreases and the entropy increases. This, for example, creates additional drag force on an aircraft with shocks.

Some kinds of shocks are:

(i) Normal shock,

(ii) Oblique shock

(iii) Bow shock

Normal shock

In the fluid mechanics, a shock wave is treated as a discontinuity where entropy increases over a nearly infinitesimal region. As no fluid flow is discontinuous, a control volume is established around the shock wave, with the control surfaces that bound this volume parallel to the shock wave. The two surfaces are separated by a very small depth such that the shock itself is entirely contained between them. At such control surfaces, momentum, mass flux, and energy are constant. It is assumed that the system is adiabatic (no heat exists or enters the system) and no work is being done.

Oblique shock

When analyzing shock waves in a flow field, which are still attached to the body, the shock wave which is deviating at some arbitrary angle from the flow direction is known as the oblique shock. These shocks require a component vector analysis of the flow. For doing so, we can treat the flow in an orthogonal direction to the oblique shock as normal chock.

Bow shock

When an oblique shock is likely to form at an angle which cannot remain on the surface, a nonlinear phenomenon arises where shock wave will form a continuous pattern around the body. These are known as the bow shocks. In these cases, the one dimensional flow model is not valid and a complex analysis is required to predict the pressure forces which are exerted on the surface.

13. Multiple choice questions

1. A tuning fork A produces 6 beats per second with the tuning fork B whose frequency is 350 Hz and 8 beats per second with the tuning fork C whose frequency is 364 Hz. The frequency of the tuning fork A is maximum at

    (i) 344 Hz   (ii) 356 Hz   (iii) 364 Hz   (iv) 372 Hz

    Ans. (ii)

2. Two tuning forks A and B produce 6 beats per second. The frequency of A is 356 Hz. The tuning fork B is filed. Two tuning

forks A and B again produced 6 beats per second. The original frequency of B is

(i) 350 Hz   (ii) 356 Hz   (iii) 362 Hz   (iv) 368 Hz

Ans. (i)

3. Two tuning forks A and B produce 6 beats per second. The frequency of A is 356 Hz. The tuning fork B is loaded with wax. Two tuning forks A and B again produced 6 beats per second. The original frequency of B is

(i) 350 Hz   (ii) 356 Hz   (iii) 362 Hz   (iv) 368 Hz

Ans. (iii)

4. For air, the pressure, temperature, volume and density are P, T, V, ρ, respectively. The expression for the velocity of sound u through the air is

(i) $u = \sqrt{\dfrac{P}{\rho}}$ (ii) $u = \sqrt{\dfrac{T}{\rho}}$ (iii) $u = \sqrt{\dfrac{V}{\rho}}$ (iv) $u = \sqrt{\dfrac{PVC}{T}}$

Ans. (i)

5. For a medium, the pressure, temperature, volume, density and elasticity are P, T, V, ρ, E respectively. The expression for the velocity of sound u through the medium is

(i) $u = \sqrt{\dfrac{P}{\rho}}$ (ii) $u = \sqrt{\dfrac{T}{\rho}}$ (iii) $u = \sqrt{\dfrac{V}{\rho}}$ (iv) $u = \sqrt{\dfrac{PVC}{T}}$

Ans. (ii)

6. For the given mass of air, the velocity of sound in air varies with the change of

(i) Pressure   (ii) Density   (iii) Temperature   (iv) Volume

Ans. (iii)

7. Under which situation, the apparent frequency will appear increased relative to the original frequency of the source.

(i) Observer is stationary and the source is moving away from the observer.

(ii) Observer is stationary and the source is coming close to observer.

(iii) Source is stationary and the observer is moving towards the observer.

(iv) Source and observer are moving towards each other.

Ans. (i)

8. Under which situation, the apparent frequency will appear increased relative to the original frequency of the source.

(i) Observer is stationary and the source is coming close to observer.

(ii) Observer is stationary and the source is moving away from the observer.

(iii) Source is stationary and the observer is moving away from the observer.

(iv) Source and observer are moving away from each other.

Ans. (i)

9. Mach number of a supersonic wave is

(i) 0   (ii) between 0 and 1   (iii) 1   (iv) more than 1

Ans. (iv)

14. Problems and questions

1. Two sound waves having frequencies n_1 and n_2 are produced in a common medium. Show that interference of the waves produces maximum and minimum alternatively after equal intervals of time $1/2(n_1 - n_2)$.

2. Two sound waves having frequencies n_1 and n_2 are produced in a common medium. Show that the number of beat produced per second is $(n_1 - n_2)$.

3. Derive the expression for the velocity of sound propagating through a medium.

4. Derive Newton expression for the velocity of sound propagating through air. Describe the correction suggested by Laplace.

5. Describe the effect of temperature and pressure on the velocity of sound in air.

6. Show that the Doppler effect is symmetric in case of light whereas in case of sound waves the Doppler effect is asymmetric.

7. Describe the Bel and decibel for the intensity of sound.

8. Write notes on the following

   (i) Speed of sound

   (ii) Shock waves

   (iii) Beats

   (iv) Dependence on temperature of the velocity of sound in air

   (v) Dependence on pressure of the velocity of sound in air

   (vi) Intensity level of a sound wave

   (vii) Shock waves

   (viii) Mach number

Appendices

A. Some physical parameters

c	Speed of light	3×10^8 m/s
e	Electron charge	1.60×10^{-19} C
m_e	Electron rest mass	9.11×10^{-31} kg
m_p	Proton rest mass	1.67×10^{-27} kg
m_p/m_e		1837
k	Boltzmann constant	1.38×10^{-23} J/K
h	Planck constant	6.62×10^{-34} J s
R	Gas constant	8.31 J/mol K
σ	Stefan constant	5.67×10^{-8} J/m^2 s K^4
N	Avogadro number	6.022×10^{26} particles/kg mol
ϵ_0	Permittivity of free space	8.854×10^{-12} F/m
μ_0	Permeability of free space	$4\pi \times 10^{-7}$ H/m

B. Useful relations

$1 \text{ eV} = 1.60 \times 10^{-19}$ J

$1 \text{ cal} = 4.18$ J

$1 \text{ amu} = 931.5$ MeV/c^2

$1 \text{ MeV/c}^2 = 1.073 \times 10^{-3}$ amu $= 1.783 \times 10^{-30}$ kg

$1 \text{ Å} = 10^{-10}$ m $= 0.1$ nm

$1 \text{ fm} = 10^{-15}$ m

$1 \text{ in} = 2.540$ cm

$hc = 1.986 \times 10^{-25}$ J m

$1 \text{ barn} = 10^{-28}$ m^2

$1 \text{ curie} = 3.7 \times 10^{10}$ decays/s

$1 \text{ watt} = 1$ J/s

C. Del operator

Here, we shall mention in brief about the operator ∇, called the del operator. In science, we generally deal with the Cartesian coordinates (x, y, z) and spherical polar coordinates (r, θ, ϕ). In the present discussion, we shall use Cartesian coordinates only. In the Cartesian coordinates (x, y, z), the operator ∇ is expressed as

$$\nabla \equiv \frac{\partial}{\partial x}\,\hat{i} + \frac{\partial}{\partial y}\,\hat{j} + \frac{\partial}{\partial z}\,\hat{k}$$

where $\hat{i}$, $\hat{j}$, $\hat{k}$ are the unit vectors along x, y and z axes, respectively. The del operator is obviously a vector-operator.

C.1. Gradient

The gradient of a scalar quantity ϕ is expressed as

$$\nabla\phi = \frac{\partial\phi}{\partial x}\,\hat{i} + \frac{\partial\phi}{\partial y}\,\hat{j} + \frac{\partial\phi}{\partial z}\,\hat{k}$$

Thus, gradient of a scalar quantity is a vector quantity.

C. 2. Divergence

The divergence of a vector quantity $\vec{A} = A_x\hat{i} + A_y\hat{j} + A_z\hat{k}$ is expressed as

$$\nabla\cdot\vec{A} = \frac{\partial A_x}{\partial x} + \frac{\partial A_y}{\partial y} + \frac{\partial A_z}{\partial z}$$

Thus, divergence of a vector quantity is a scalar quantity.

C.3. Curl

The curl of a vector quantity $\vec{A} = A_x\hat{i} + A_y\hat{j} + A_z\hat{k}$ is expressed as

$$\nabla\times\vec{A} = \begin{vmatrix} \hat{i} & \hat{j} & \hat{k} \\ \frac{\partial}{\partial x} & \frac{\partial}{\partial y} & \frac{\partial}{\partial z} \\ A_x & A_y & A_z \end{vmatrix}$$

$$= \left(\frac{\partial A_z}{\partial y} - \frac{\partial A_y}{\partial z}\right)\hat{i} + \left(\frac{\partial A_x}{\partial z} - \frac{\partial A_z}{\partial x}\right)\hat{j} + \left(\frac{\partial A_y}{\partial x} - \frac{\partial A_x}{\partial y}\right)\hat{k}.$$

Exercise 1: We are given a scalar quantity $\phi = x^2y + y^2z + z^2x$ and a vector quantity $\vec{A} = x^2y\,\hat{i} + y^2z\,\hat{j} + z^2x\,\hat{k}$. Calculate (i) $\nabla\phi$, (ii) $\nabla\cdot\vec{A}$ and (iii) $\nabla\times\vec{A}$.

Solution: (i) We have

$$\nabla\phi = \frac{\partial\phi}{\partial x}\,\hat{i} + \frac{\partial\phi}{\partial y}\,\hat{j} + \frac{\partial\phi}{\partial z}\,\hat{k}$$

$$= (2xy + z^2)\hat{i} + (2yz + x^2)\hat{j} + (2zx + y^2)\hat{k}$$

(ii) We have

$$\nabla\cdot\vec{A} = \frac{\partial A_x}{\partial x} + \frac{\partial A_y}{\partial y} + \frac{\partial A_z}{\partial z}$$

$$= \frac{\partial(x^2y)}{\partial x} + \frac{\partial(y^2z)}{\partial y} + \frac{\partial z^2x}{\partial z} = 2xy + 2yz + 2zx$$

(iii) We have

$$\nabla \times \vec{A} = \begin{vmatrix} \hat{i} & \hat{j} & \hat{k} \\ \frac{\partial}{\partial x} & \frac{\partial}{\partial y} & \frac{\partial}{\partial z} \\ A_x & A_y & A_z \end{vmatrix} = \begin{vmatrix} \hat{i} & \hat{j} & \hat{k} \\ \frac{\partial}{\partial x} & \frac{\partial}{\partial y} & \frac{\partial}{\partial z} \\ x^2 y & y^2 z & z^2 x \end{vmatrix}$$

$$= \left[\frac{\partial(z^2 x)}{\partial y} - \frac{\partial(y^2 z)}{\partial z}\right]\hat{i} + \left[\frac{\partial(x^2 y)}{\partial z} - \frac{\partial(z^2 x)}{\partial x}\right]\hat{j} + \left[\frac{\partial(y^2 z)}{\partial x} - \frac{\partial(x^2 y)}{\partial y}\right]\hat{k}$$

$$= -y^2 \hat{i} - z^2 \hat{j} - x^2 \hat{k}$$

Exercise 2: For a scalar quantity ϕ and a vector quantity $\vec{A}$ show that (i) $\nabla \times (\nabla \phi) = 0$ and (ii) $\nabla \cdot (\nabla \times \vec{A}) = 0$.

Solution: (i) We have

$$\nabla \phi = \frac{\partial \phi}{\partial x}\,\hat{i} + \frac{\partial \phi}{\partial y}\,\hat{j} + \frac{\partial \phi}{\partial z}\,\hat{k}$$

and

$$\nabla \times (\nabla \phi) = \begin{vmatrix} \hat{i} & \hat{j} & \hat{k} \\ \frac{\partial}{\partial x} & \frac{\partial}{\partial y} & \frac{\partial}{\partial z} \\ \frac{\partial \phi}{\partial x} & \frac{\partial \phi}{\partial y} & \frac{\partial \phi}{\partial z} \end{vmatrix}$$

$$= \left(\frac{\partial^2 \phi}{\partial y \partial z} - \frac{\partial^2 \phi}{\partial z \partial y}\right)\hat{i} - \left(\frac{\partial^2 \phi}{\partial x \partial z} - \frac{\partial^2 \phi}{\partial z \partial x}\right)\hat{j} + \left(\frac{\partial^2 \phi}{\partial x \partial y} - \frac{\partial^2 \phi}{\partial y \partial x}\right)\hat{k} = 0$$

(ii) We have

$$\nabla \times \vec{A} = \begin{vmatrix} \hat{i} & \hat{j} & \hat{k} \\ \frac{\partial}{\partial x} & \frac{\partial}{\partial y} & \frac{\partial}{\partial z} \\ A_x & A_y & A_z \end{vmatrix}$$

$$= \left(\frac{\partial A_z}{\partial y} - \frac{\partial A_y}{\partial z}\right)\hat{i} - \left(\frac{\partial A_z}{\partial x} - \frac{\partial A_x}{\partial z}\right)\hat{j} + \left(\frac{\partial A_y}{\partial x} - \frac{\partial A_x}{\partial y}\right)\hat{k}$$

and

$$\nabla \cdot (\nabla \times \vec{A}) = \frac{\partial}{\partial x}\left(\frac{\partial A_z}{\partial y} - \frac{\partial A_y}{\partial z}\right) - \frac{\partial}{\partial y}\left(\frac{\partial A_z}{\partial x} - \frac{\partial A_x}{\partial z}\right) + \frac{\partial}{\partial z}\left(\frac{\partial A_y}{\partial x} - \frac{\partial A_x}{\partial y}\right)$$

$$= \left(\frac{\partial^2 A_z}{\partial x \partial y} - \frac{\partial^2 A_y}{\partial x \partial z}\right) - \left(\frac{\partial^2 A_z}{\partial y \partial x} - \frac{\partial^2 A_x}{\partial y \partial z}\right) + \left(\frac{\partial^2 A_y}{\partial z \partial x} - \frac{\partial^2 A_x}{\partial z \partial y}\right) = 0$$

D. Gauss theorem

This theorem relates an integral over a volume to an integral over a surface, enclosing the volume and is expressed as

$$\int_V (\nabla \cdot \vec{F})\, dV = \int_S \vec{F} \cdot d\vec{S}$$

Here, we have volume element dV and surface element $d\vec{S}$. Total volume on the left side of equation is that which is bounded by total surface area on the right side of equation. Obviously, we have a close surface.

E. Stokes theorem

This theorem relates an integral over a surface to an integral over an integral over a line, enclosing the surface, and is expressed as

$$\int_S (\nabla \times \vec{F}) \cdot d\vec{S} = \int_C \vec{F} \cdot d\vec{l}$$

Here, we have surface element $d\vec{S}$ and line element $d\vec{l}$. Total surface area on the left side of equation is that which is bounded by the total length on the right side of equation. Obviously, we have an open surface.

Bibliography

G. Aruldhas (2012) Engineering Physics, Prentice-Hall of India Pvt. Ltd., New Delhi

A. Beiser, S. Mahajan & S. Rai Choudhary (2009) Concepts of Modern Physics, Tata McGraw Hill Education Pvt. Ltd., New Delhi

F.S. Crawford (2011) Waves, Tata McGraw Hill Education Pvt. Ltd., New Delhi

A.K. Jha (2009) A Textbook of Applied Physics, I.K. International Publishing House, Pvt. Ltd., New Delhi.

F.A. Jenkins & H.E. White (2011) Fundamentals of Optics, IV Edition, Tata McGraw Hill Education Pvt. Ltd., New Delhi

S.L. Kakani & S. Kakani (2008) Engineering Physics, CBS Publishers & Distributors, New Delhi

H.K. Malik & A.K. Singh (2010) Engineering Physics, Tata McGraw Hill Education Pvt. Ltd., New Delhi

B.K. Pandey & S. Chaturvedi (2012) Engineering Physics, Cengage Learning India Pvt. Ltd., New Delhi

D. Singh (2010) Fundamentals of Optics, PHI Learning Pvt. Ltd., New Delhi

N. Subrahmanyam & B. Lal (2012) Waves and Oscillations, Vika Publishing House Pvt. Ltd., Noida

N. Subrahmanyam, B. Lal & M.N. Avadhanulu (2010) A Textbook of Optics, S. Chand & Company Ltd., New Delhi

Index